Lectures on Algebraic Cycles

Second Edition

Spencer Bloch's 1979 Duke lectures, a milestone in modern mathematics, have been out of print almost since their first publication in 1980, yet they have remained influential and are still the best place to learn the guiding philosophy of algebraic cycles and motives. This edition, now professionally typeset, has a new preface by the author giving his perspective on developments in the field over the past 30 years.

The theory of algebraic cycles encompasses such central problems in mathematics as the Hodge conjecture and the Bloch–Kato conjecture on special values of zeta functions. The book begins with Mumford's example showing that the Chow group of zero-cycles on an algebraic variety can be infinite dimensional, and explains how Hodge theory and algebraic K-theory give new insights into this and other phenomena.

SPENCER BLOCH is R. M. Hutchins Distinguished Service Professor in the Department of Mathematics at the University of Chicago.

NEW MATHEMATICAL MONOGRAPHS

All the titles listed below can be obtained from good booksellers or from Cambridge University Press. For a complete series listing visit http://www.cambridge.org/uk/series/sSeries.asp?code=NMM

Lectures on Algebraic Cycles
Second Edition

SPENCER BLOCH
University of Chicago

CAMBRIDGE
UNIVERSITY PRESS

CAMBRIDGE
UNIVERSITY PRESS

32 Avenue of the Americas, New York NY 10013-2473, USA

It furthers the University's mission by disseminating knowledge in the pursuit of
education, learning and research at the highest international levels of excellence.

www.cambridge.org
Information on this title: www.cambridge.org/9780521118422

© S. Bloch 2010
© Mathematics Department, Duke University 1980

First edition published in 1980 by Duke University, Durham, NC 27706, USA
Second edition published 2010

A catalogue record for this publication is available from the British Library

ISBN 978-0-521-11842-2 Hardback

Contents

Preface to the second edition

30 Years later...

Looking back over these lectures, given at Duke University in 1979, I can say with some pride that they contain early hints of a number of important themes in modern arithmetic geometry. Of course, the flip side of that coin is that they are now, thirty years later, seriously out of date. To bring them up to date would involve writing several more monographs, a task best left to mathematicians thirty years younger than me. What I propose instead is to comment fairly briefly on several of the lectures in an attempt to put the reader in touch with what I believe are the most important modern ideas in these areas. The section on motives just below is intended as a brief introduction to the modern viewpoint on that subject. The remaining sections until the last follow roughly the content of the original book, though the titles have changed slightly to reflect my current emphasis. The last section, motives in physics, represents my recent research.

In the original volume I included a quote from Charlie Chan, the great Chinese detective, who told his bumbling number one son "answer simple, but question very very hard." It seemed to me an appropriate comment on the subject of algebraic cycles. Given the amazing deep new ideas introduced into the subject in recent years, however, I think now that the question remains very very hard, but the answer is perhaps no longer so simple...

At the end of this essay I include a brief bibliography, which is by no means complete. It is only intended to illustrate the various ideas mentioned in the text.

Motives

Much of the recent work in this area is centered around motives and the construction – in fact various constructions, due to Hanamura (1995), Levine (1998), and Voevodsky (Mazza et al. 2006; Voevodsky et al. 2000) – of a triangulated category of mixed motives. I will sketch Voevodsky's construction as it also plays a central role in his proof of the Bloch–Kato–Milnor conjecture discussed in Lecture 5. Then I will discuss various lectures from the original monograph.

Let k be a field. The category Cor_k is an additive category with objects smooth k-varieties: $Ob(Cor_k) = Ob(Sm_k)$. Morphisms $Z\colon X \to Y$ are finite linear combinations of correspondences $Z = \sum n_i Z_i$ where $Z_i \subset X \times Y$ is closed and the projection $\pi_i\colon Z_i \to X$ is finite and surjective. Intuitively, we may think of Z_i as a map $X \to \operatorname{Sym}^n Y$ associating to $x \in X$ the fibre $f_i^{-1}(x)$ viewed as a zero-cycle on Y. There is an evident functor

$$Sm_k \to Cor_k$$

which is the identity on objects.

A presheaf on a category C with values in an abelian category $\mathcal{A}$ is simply a contravariant functor $F\colon C^{\mathrm{op}} \to \mathcal{A}$. An $\mathcal{A}$-valued presheaf F on Cor_k induces a presheaf $F|_{Sm_k}$ on Sm_k. Intuitively, to lift a presheaf G from Sm_k to Cor_k one needs a structure of trace maps or transfers $f_*\colon G(Z) \to G(X)$ for Z/X finite. Presheaves on Cor_k are referred to as *presheaves with transfers*.

For $X \in Ob(Sm_k)$ one has the representable sheaf $\mathbf{Z}_{\mathrm{tr}}(X)$ defined by

$$\mathbf{Z}_{\mathrm{tr}}(X)(U) = \operatorname{Hom}_{Cor_k}(U, X).$$

An important elaboration on this idea yields for pointed objects $x_i \in X_i$

$$\mathbf{Z}_{\mathrm{tr}}((X_1, x_1) \wedge \cdots \wedge (X_n, x_i))$$
$$:= \operatorname{Coker}\!\left(\bigoplus \mathbf{Z}_{\mathrm{tr}}(X_1 \times \cdots \widehat{X_i} \times \cdots X_n) \to \mathbf{Z}_{\mathrm{tr}}(\textstyle\prod X_i)\right).$$

In particular, one defines $\mathbf{Z}_{\mathrm{tr}}(\bigwedge^n \mathbf{G}_{\mathrm{m}})$ by taking $X_i = \mathbf{A}^1 - \{0\}$ and $x_i = 1$.

A presheaf with transfers F is called *homotopy invariant* if, with obvious notation, $i_0^* = i_1^*\colon F(U \times \mathbf{A}^1) \to F(U)$. The complex of *chains* $C_*(F)$ on a presheaf with transfers F is the presheaf of complexes (placed in cohomological degrees $[-\infty, 0]$)

$$C_*(F) := U \mapsto \cdots \to F(U \times \Delta^n) \to \cdots \to F(U \times \Delta^0)$$

Here

(0.1) $$\Delta^n := \operatorname{Spec} k[t_0, \ldots, t_n]/(\textstyle\sum t_i - 1)$$

is the algebro-geometric n-simplex. The boundary maps in the complex are the usual alternating sums of restrictions to the faces $\Delta^{n-1} \hookrightarrow \Delta^n$ defined by setting $t_i = 0$. The two restrictions

$$i_0^*, i_1^* : C_*(F)(U \times \mathbf{A}^1) \to C_*(F)(U)$$

are shown to be homotopic, so the homology presheaves $H_n(C_*(F))$ are homotopy invariant.

Maps $f_0, f_1 : X \to Y$ in Cor_k are $\mathbf{A}^1$-*homotopic* if there exists $H : X \times \mathbf{A}^1 \to Y$ in Cor_k such that $f_j = i_j^* H$. $\mathbf{A}^1$-homotopy is an equivalence relation, and $\mathbf{A}^1$-homotopic maps induce homotopic maps

$$f_{0*} \simeq f_{1*} : C_* \mathbf{Z}_{\mathrm{tr}}(X) \to C_* \mathbf{Z}_{\mathrm{tr}}(Y).$$

Voevodsky defines

$$\mathbf{Z}(q) := C_* \mathbf{Z}_{\mathrm{tr}}(\textstyle\bigwedge^q \mathbf{G}_{\mathrm{m}})[-q], \quad q \geq 0.$$

More precisely, the above complex is viewed as a complex of presheaves on Sm_k and then localized for the Zariski topology. Motivic cohomology is then defined (for $q \geq 0$) as the hypercohomology of this complex of Zariski sheaves:

$$H_{\mathrm{M}}^p(X, \mathbf{Z}(q)) := H_{\mathrm{Zar}}^p(X, \mathbf{Z}(q)).$$

One has a notion of tensor product for presheaves on the category Cor_k, and $\mathbf{Z}_{\mathrm{tr}}(X) \otimes \mathbf{Z}_{\mathrm{tr}}(Y) = \mathbf{Z}_{\mathrm{tr}}(X \times Y)$. In particular, $\mathbf{Z}(p) \otimes \mathbf{Z}(q) \to \mathbf{Z}(p+q)$ so one gets a product structure on motivic cohomology. In low degrees one has

$$H_{\mathrm{M}}^0(X, \mathbf{Z}(0)) = \mathbf{Z}[\pi_0(X)],$$

$$H_{\mathrm{M}}^p(X, \mathbf{Z}(0)) = (0), \quad p > 0,$$

$$\mathbf{Z}(1) \cong \mathbf{G}_{\mathrm{m}}[-1].$$

Another important sign that this is the right theory is the link with Milnor K-theory. ($K_*^{\mathrm{Milnor}}(k)$ is defined as the quotient of the tensor algebra on $k^\times$ by the ideal generated by quadratic relations $a \otimes (1 - a)$ for $a \in k - \{0, 1\}$.)

Theorem $H_{\mathrm{M}}^n(\operatorname{Spec} k, \mathbf{Z}(n)) \cong K_n^{\mathrm{Milnor}}(k).$

The fact that the Zariski topology suffices to define motivic cohomology is somewhat surprising because a Zariski open cover $\pi : U \to X$ does not yield a resolution of Zariski sheaves

(0.2) $$\cdots \to \mathbf{Z}_{\mathrm{tr}}(U \times_X U) \to \mathbf{Z}_{\mathrm{tr}}(U) \to \mathbf{Z}(X) \to 0.$$

To remedy this, Voevodsky employs the Nisnevich topology. A morphism $\pi : U \to X$ is a Nisnevich cover if for any field K/k one has $U(K) \twoheadrightarrow X(K)$.

To see that (0.2) becomes exact when localized for the Nisnevich topology, one uses the fact that any finite cover of a Hensel local ring is a product of local rings.

The actual triangulated category of effective motives over k is a quotient category of the derived category $D^-(Sh_{\mathrm{Nis}}(Cor_k))$ of bounded-below complexes of Nisnevich sheaves on Cor_k. One considers the smallest thick subcategory W containing all cones of $\mathbf{Z}_{\mathrm{tr}}(X \times \mathbf{A}^1) \to \mathbf{Z}_{\mathrm{tr}}(X)$, and one defines

$$DM_{\mathrm{Nis}}^{\mathrm{eff}}(k) := D^- Sh_{\mathrm{Nis}}(Cor_k)[W^{-1}].$$

Said another way, one formally inverts all morphisms with cones in W. Finally, the motive associated to a smooth k-variety X is defined by

$$(0.3) \qquad\qquad M(X) := \mathbf{Z}_{\mathrm{tr}}(X) \in DM_{\mathrm{Nis}}^{\mathrm{eff}}(k).$$

The category of *geometric* motives $DM_{\mathrm{geo}}^{\mathrm{eff}}(k)$ is the thick subcategory in $DM_{\mathrm{Nis}}^{\mathrm{eff}}(k)$ generated by the $M(X)$.

One has the following properties:

Mayer-Vietoris

$$M(U \cap V) \to M(U) \oplus M(V) \to M(X) \to M(U \cap V)[1]$$

is a distinguished triangle.

Künneth

$$M(X \times Y) = M(X) \otimes M(Y).$$

Vector bundle theorem

$$M(X) \cong M(V)$$

for V/X a vector bundle.

Cancellation   Assume varieties over k admit a resolution of singularities. Write $M(q) := M \otimes \mathbf{Z}(q)$. Then

$$\mathrm{Hom}(M, N) \cong \mathrm{Hom}(M(q), N(q)).$$

The category of (not necessarily effective) motives is obtained by inverting the functor $M \mapsto M(1)$ in DM^{eff}.

Projective bundle theorem   For V/X a rank $n + 1$ vector bundle

$$M(\mathbf{P}(V)) \cong \bigoplus_{i=0}^{n} M(X)(i)[2i].$$

Chow motives If X, Y are smooth projective, then

$$(0.4) \qquad \operatorname{Hom}(M(X), M(Y)) \cong \operatorname{CH}^{\dim X}(X \times Y).$$

The category of *Chow motives* over a field k has as objects triples (X, p, m) with X smooth projective over k, $p \in \operatorname{CH}^{\dim X}(X \times X)_{\mathbf{Q}}$ a projector (i.e. $p \circ p = p$) and $m \in \mathbf{Z}$. The morphisms are given by

$$(0.5) \qquad \operatorname{Hom}((X, p, m), (Y, q, n)) := q \circ \operatorname{CH}^{\dim X + n - m}(X \times Y) \circ p.$$

It follows from (0.4) and the existence of projectors in $DM_{\mathrm{Nis}}^{\mathrm{eff}}(k)$ that the category of Chow motives embeds in $DM_{\mathrm{Nis}}^{\mathrm{eff}}(k)$.

Motivic cohomology For X/k smooth, we have

$$(0.6) \qquad H_M^p(X, \mathbf{Z}(q)) \cong \operatorname{Hom}_{DM_{\mathrm{Nis}}^{\mathrm{eff}}(k)}(\mathbf{Z}_{\mathrm{tr}}(X), \mathbf{Z}(q)[p]).$$

In fact, motivic cohomology is closely related to algebraic cycles, and this relationship lies at the heart of modern cycle theory. There are a number of ways to formulate things. I will use *higher Chow groups* because they relate most naturally to arithmetic questions. Let $\Delta^{\bullet}$ be the cosimplicial variety as in (0.1) above. Define $\mathcal{Z}^q(X, n)$ to be the free abelian group of algebraic cycles on $X \times \Delta^n$ which are in good position with respect to all faces $X \times \Delta^m \hookrightarrow X \times \Delta^n$. The complex $\mathcal{Z}^q(X, \bullet)$ is defined by taking alternating sums of pullbacks in the usual way:

$$(0.7) \qquad \cdots \to \mathcal{Z}^q(X, 2) \to \mathcal{Z}^q(X, 1) \to \mathcal{Z}^q(X, 0) \to 0.$$

(Here $\mathcal{Z}^q(X, n)$ is placed in cohomological degree $-n$.) The *higher Chow groups* are defined by

$$(0.8) \qquad \operatorname{CH}^q(X, n) := H^{-n}(\mathcal{Z}^q(X, \bullet)).$$

For example, the usual Chow group $\operatorname{CH}^q(X) = \operatorname{CH}^q(X, 0)$. Voevodsky proves that for X smooth over a perfect field k one has

$$(0.9) \qquad H_M^p(X, \mathbf{Z}(q)) \cong \operatorname{CH}^q(X, 2q - p) = H^p(\mathcal{Z}^q(X, \bullet)[-2q]).$$

Beilinson and Soulé conjecture that the shifted chain complex $\mathcal{Z}^q(X, \bullet)[-2q] \otimes \mathbf{Q}$ has cohomological support in degrees $[0, 2q]$. Actually, their conjecture was formulated in terms of the γ-filtration in K-theory, but one has the further identification

$$(0.10) \qquad H_M^p(X, \mathbf{Z}(q)) \otimes \mathbf{Q} \cong \operatorname{CH}^q(X, 2q - p) \otimes \mathbf{Q} \cong \operatorname{gr}_{\gamma}^q K_p(X) \otimes \mathbf{Q}.$$

Lecture 1: Zero-cycles

The two most important ideas here are firstly the conjecture that surfaces with geometric genus zero ($p_g = 0$) should have Chow group of zero-cycles representable. For S such a surface over $\mathbf{C}$ we expect an exact sequence

$$0 \to \mathrm{Alb}(S) \to \mathrm{CH}_0(S) \xrightarrow{\deg} \mathbf{Z} \to 0.$$

The group $T(S)$ defined in Lemma 1.4 is conjectured to be zero in this case. Secondly, for any smooth projective variety X, the Chow group of zero-cycles $\mathrm{CH}_0(X)$ is conjectured (1.8) to carry a descending filtration $F^*\mathrm{CH}_0(X)$ which is functorial for correspondences such that the map $\mathrm{gr}_F^p\mathrm{CH}_0(X) \xrightarrow{\Lambda} \mathrm{gr}_F^p\mathrm{CH}_0(Y)$ induced by an algebraic cycle $\Lambda \in \mathrm{CH}^{\dim Y}(X \times Y)$ depends only on the class of Λ in cohomology. Indeed, one may conjecture the existence of such a filtration on $\mathrm{CH}_q(X)$ for any q.

These conjectures remain unproven, but a very beautiful general picture, based on the yoga of mixed motives, has been elaborated by A. Beilinson. Interested readers should consult the important article by Jannsen (1994) and the literature cited there. Let me sketch briefly (following Jannsen) the modern viewpoint.

It is convenient to dualize the definition of $M(X)$ (0.3). Assume X smooth, projective of dimension d. Define (<u>Hom</u> means the internal Hom in DM)

$$M(X)^* = \underline{\mathrm{Hom}}_{DM}(M(X), \mathbf{Z}(0)).$$

The formula for the Chow group becomes

$$\mathrm{CH}^p(X) = H_M^{2p}(X, \mathbf{Z}(p)) = \mathrm{Hom}_{DM}(\mathbf{Z}(0), M(X)^*(p)[2p]).$$

One of Grothendieck's *standard conjectures* about algebraic cycles is that there exist *Künneth projectors*

$$\pi_i \in \mathrm{CH}^d(X \times X)_{\mathbf{Q}}/\{\text{homological equivalence}\}$$

inducing the natural projections $H^*(X) \to H^i(X)$ on cohomology. If we assume further that the ideal $\{\text{homological equiv.}\}/\{\text{rational equiv.}\} \subset \mathrm{CH}^d(X \times X)_{\mathbf{Q}}$ is nilpotent (nilpotence conjecture), then the pr_i lift (non-canonically) to projectors $\mathrm{pr}_{i,\mathrm{rat}} \in \mathrm{CH}^d(X \times X)_{\mathbf{Q}}$ and we may use (0.5) to decompose $M(X)^* \otimes \mathbf{Q} = \bigoplus_i h^i[-i]$ non-canonically as a direct sum of Chow motives. This idea is due to J. Murre (1993a). The hope is that DM admits a t-structure such that

$$(0.11) \qquad\qquad H^i(M(X)^* \otimes \mathbf{Q}) = h^i(X).$$

The resulting spectral sequence

$$E_2^{p,q} = \mathrm{Hom}_{DM}(\mathbf{Z}(0), H^q(M(X)^*(j)[p]) \Rightarrow \mathrm{Hom}_{DM}(\mathbf{Z}(0), M^*(X)(j)[p+q])$$

would yield filtrations on the Chow groups $\otimes \mathbf{Q}$ with

$$F^\nu \mathrm{CH}^j(X)_{\mathbf{Q}} \cong \bigoplus_{i=0}^{2j-\nu} \mathrm{Ext}_{DM}^{2j-i}(\mathbf{Q}(0), h^i(X)(j)),$$

$$\mathrm{gr}_F^\nu \mathrm{CH}^j(X)_{\mathbf{Q}} \cong \mathrm{Ext}_{DM}^\nu(\mathbf{Q}(0), H^{2j-\nu}(X)(j)).$$

Murre suggests a natural strengthening of his conjectures, based on the idea that one should be able for $i \leq d$, to find representatives for π_i supported on $X_i \times X \subset X \times X$, where $X_i \subset X$ is a general plane section of dimension i. For example, $\pi_0 = \{x\} \times X$ for a point x. Clearly this would imply $\pi_i \mathrm{CH}^j(X) = 0$ for $i < j$, and since the conjectures imply

$$F^\nu \mathrm{CH}^j(X) = \bigoplus_{i=0}^{2j-\nu} \pi_i \, \mathrm{CH}^j(X),$$

we could conclude further that

$$F^\nu \mathrm{CH}^j(X) = (0), \quad \nu > j.$$

Suppose, for example, that $\dim X = 2$. We would get a 3-step filtration on the zero-cycles: $\mathrm{CH}_0(X) = F^0 \supset F^1 \supset F^2 \supset (0)$, with

$$\mathrm{gr}_F^0 \mathrm{CH}_0(X)_{\mathbf{Q}} = \mathrm{Hom}_{DM}(\mathbf{Q}(0), h^4(X)(2)),$$

$$\mathrm{gr}_F^1 = \mathrm{Ext}_{DM}^1(\mathbf{Q}(0), h^3(X)(2)),$$

$$\mathrm{gr}_F^2 = \mathrm{Ext}_{DM}^2(\mathbf{Q}(0), h^2(X)(2)).$$

This fits perfectly with the ideas in Lecture 1. Indeed, Murre has computed gr^0 and gr^1 and he finds exactly the degree and the Albanese. Of course, gr^2 is more problematical, but note that the condition discussed in the text that $H^2(X, \mathbf{Q}_\ell(1))$ should be generated by divisors (which is equivalent to $p_g = 0$ in characteristic zero) means $h^2(X)(2) \cong \oplus \mathbf{Q}(1)$. (This is obvious for motives modulo homological equivalence. Assuming the nilpotence conjecture, it holds for Chow motives as well.) In this case, gr_F^2 can be computed for $X = \mathbf{P}^2$ when it is clearly (0).

The conjectural "theorem of the hypersquare" (Proposition 1.12) can be understood in motivic terms (Jannsen 1994, conj. 3.12) using the fact that $h^n(X_0 \times \cdots \times X_n)$ is a direct summand of $\bigoplus h^n(X_0 \times \cdots \widehat{X_i} \cdots \times X_n)$.

Clearly wrongheaded, however, is Metaconjecture 1.10, which stated that $F^2\mathrm{CH}_0(X)$ is controlled by the polarized Hodge structure associated to $H^2(X)$. Indeed, $\mathrm{Ext}^2 = (0)$ in the category of Hodge structures. One may try (compare Carlson and Hain 1989) to look at Exts in some category of variations of Hodge structure. In the absence of parameters supporting such variations (i.e. for X over a number field), however, the Ext^2 term should vanish and we should have $F^2\text{CH}^2(X) = (0)$.

Lectures 2 and 3: Intermediate jacobians

The modern point of view about intermediate jacobians is to view them as $\text{Ext}^1(\mathbf{Z}(0), H)$ where H is a suitable Hodge structure, and the Ext group is taken in the abelian category of mixed Hodge structures (Carlson 1987). In the classic situation, $H = H^{2r-1}(X, \mathbf{Z}(r))$ where X is a smooth projective variety. Note in this case that H has weight -1. An extension $0 \to H \to E \to \mathbf{Z}(0) \to 0$ would yield a mixed Hodge structure E with weights $0, -1$ and Hodge filtration

$$E_\mathbf{C} = F^{-1}E_\mathbf{C} \supset F^0 E_\mathbf{C} \cdots .$$

Let $f, w \in E_\mathbf{C}$ be liftings of $1 \in \mathbf{Z}(0)$ splitting the weight and Hodge filtrations respectively. The difference between them $w - f$ gives a well-defined class in $J = H_\mathbf{C}/(F^0 H_\mathbf{C} + H_\mathbf{Z})$ which is the intermediate jacobian. To define the class of a codimension-r cycle $Z = \sum n_i Z_i$, let $|Z|$ be the support of the cycle. We have a cycle class with supports $[Z]: \mathbf{Z} \to H^{2r}_{|Z|}(X, \mathbf{Z}(r))$ and a diagram

(0.12)
$$H^{2r-1}_{|Z|}(X, \mathbf{Z}(r)) \longrightarrow H^{2r-1}(X, \mathbf{Z}(r)) \longrightarrow H^{2r-1}(X - |Z|, \mathbf{Z}(r))$$

$$\longrightarrow H^{2r}_{|Z|}(X, \mathbf{Z}(r)) \longrightarrow H^{2r}(X, \mathbf{Z}(r))$$
$$ \quad |Z| \uparrow \qquad\qquad \|$$
$$ \quad \mathbf{Z} \xrightarrow{\quad 0 \quad} H^{2r}(X, \mathbf{Z}(r))$$

The first group $H^{2r-1}_{|Z|}(X, \mathbf{Z}(r))$ is zero by purity, and vanishing of the lower right-hand arrow will hold if Z is homologous to 0. Assuming this, we get the desired extension of Hodge structures. (I believe this construction is due to Deligne, though I do not have a good reference.)

For H any mixed Hodge structure with weights < 0 one has an analogous construction. Note, however, the resulting abelian Lie group J need not be compact. For example, $\text{Ext}^1(\mathbf{Z}(0), \mathbf{Z}(1)) \cong \mathbf{C}^\times \cong S^1 \times \mathbf{R}$. In Lecture 3, the focus is on the case

$$H = H^1(C, \mathbf{Z}(2)) \otimes H^1_c(\mathbf{G}_\mathrm{m}, \mathbf{Z})^{\otimes 2} \subset H^3_c(C \times (\mathbf{G}_\mathrm{m})^2, \mathbf{Z}(2)).$$

Here H has weight -3, and

$$\text{Ext}^1_{MHS}(\mathbf{Z}(0), H) \cong H^1(C, \mathbf{C})/H^1(C, \mathbf{Z}(2)) \cong H^1(C, \mathbf{C}^\times(1)).$$

At the time I was very much inspired by the work of Borel (1977) on regulators for higher K-groups of number fields. I believed that similar regulators could be defined for arithmetic algebraic varieties more generally, and that these regulators could be related to values of Hasse–Weil L-functions. This was done in a very limited and ad hoc way in Bloch (1980, 2000), and then much more definitively by Beilinson (1985). From the point of view of Lecture 3, the regulator can be thought of as a relative cycle class map

$$H^p_M(X, \mathbf{Z}(q)) \overset{(0.9)}{\cong} \text{CH}^q(X, 2q-p) \to \text{Ext}^1_{MHS}(\mathbf{Z}(0), H) = H_\mathbf{C}/(H_\mathbf{Z}+F^0 H_\mathbf{C}).$$

Here $H = H^{2q-1}(X \times \Delta^{2q-p}, X \times \partial\Delta^{2q-p}; \mathbf{Z}(q))$. For details of this construction, see Bloch (2000). Other constructions are given in Bloch (1986b) and Goncharov (2005).

The quotient of this Ext group by its maximal compact subgroup is the corresponding Ext in the category of **R**-Hodge structures. It is an **R**-vector space. More generally one can associate to any mixed Hodge structure a nilpotent matrix γ (Cattani et al., 1986, prop. 2.20), which is the obstruction to a real splitting of the filtration by weights. These invariants arise, for example, if the curve C in Lecture 3 is allowed to degenerate, so $H^1(C, \mathbf{Z})$ is itself a mixed Hodge structure.

Lecture 4: Cohomological methods

This chapter contains basic information about algebraic K-theory, an important tool in the study of algebraic cycles. I describe the "Quillen trick" and use it to construct the Gersten resolution for K-sheaves and also Betti and étale cohomology sheaves for smooth varieties. Briefly, one considers Zariski sheaves $\mathcal{K}_q$ (resp. $\mathcal{H}^q_{\text{Betti}}$, resp. $\mathcal{H}^q_{\text{et}}$) associated to the presheaf $U \mapsto K_q(U)$ of algebraic K-groups (resp. $U \mapsto H^q_{\text{Betti}}(U, \mathbf{Z})$, resp. $U \mapsto H^q_{\text{et}}(U, \mathbf{Z}/n\mathbf{Z})$). One obtains resolutions of these sheaves which enable one to identify, for example,

(0.13)
$$H^p(X, \mathcal{K}_p) \cong \text{CH}^p(X),$$

$$H^p(X, \mathcal{H}^p_{\text{Betti}}) \cong \text{CH}^p(X)/\{\text{algebraic equivalence}\}.$$

I think it is fair to say that the resulting K-cohomology and the parallel constructions for Betti and étale cohomology have had important technical applications but have not been the breakthrough one had hoped for at the time.

Despite the Gersten resolution, it turns out to be difficult to interpret the resulting cohomology. Finiteness results, for example, are totally lacking. One nice application of the Betti theory (see reference [6] at the end of Lecture 4) was to falsify a longstanding conjecture about differential forms of the second kind on varieties of dimension ≥ 3. The spectral sequence $E_2^{p,q} = H_{\mathrm{Zar}}^p(X, \mathcal{H}_{\mathrm{Betti}}^q) \to H_{\mathrm{Betti}}^{p+q}(X, \mathbf{Z})$ leads to an exact sequence

$$(0.14) \qquad H_{\mathrm{Betti}}^3(X) \xrightarrow{a} \Gamma(X, \mathcal{H}^3) \xrightarrow{b} H^2(X, \mathcal{H}^2) \xrightarrow{c} H_{\mathrm{Betti}}^4(X)$$

Using (0.13) one can identify $\mathrm{Ker}(c)$ in (0.14) with the Griffiths group of cycles homologous to zero modulo algebraic equivalence, a group which in some cases is known not to be finitely generated (Clemens 1983). It follows in such cases that a is not surjective, indeed $\mathrm{Coker}(a)$ is infinitely generated. But $\Gamma(X, \mathcal{H}^3)$ is precisely the space of meromorphic 3-forms of the second kind; that is meromorphic forms which at every point differ from an algebraic form which is regular at the point by an exact algebraic form. Thus, unlike the case of curves and surfaces, differential forms of the second kind do not necessarily come from global cohomology classes in dimensions ≥ 3.

Lecture 5: The conjecture of Milnor–Bloch–Kato

Let F be a field and ℓ a prime with $1/\ell \in K$. The Milnor ring $K_*^{\mathrm{M}}(F)/\ell$ is generated by $F^\times/F^{\times \ell}$ with relations given by Steinberg symbols $f \otimes (1-f)$, $f \neq 0, 1$. The conjecture in question states that the natural map to Galois cohomology

$$K_*^{\mathrm{M}}(F)/\ell \to H^*(F, \mu_\ell^*)$$

is an isomorphism. My own contribution to this, which is explained in Lecture 5, is a proof that $K_n^{\mathrm{M}}(F) \to H^n(F, \mu_\ell^{\otimes n})$ is surjective when F has cohomological dimension n. For some years Voevodsky has been working on a very difficult program, using his own motivic theory and results of M. Rost, to prove the conjecture in complete generality. The proof is now complete (for an outline with references, see the webpage of C. Weibel), but there is still no unified treatment and the arguments use sophisticated techniques in algebraic homotopy theory which I have not understood.

Geometrically, the result can be formulated as follows. Let $i: X \hookrightarrow Y$ be a closed immersion of varieties over a field k of characteristic prime to ℓ, and let $j: Y - X \hookrightarrow Y$ be the open immersion. For simplicity I assume $\mu_\ell \subset k$ so there is no need to distinguish powers of μ_ℓ. Consider the exact sequence

$$H^p(Y, \mathbf{Z}/\ell) \xrightarrow{i^*} H^p(X, \mathbf{Z}/\ell) \xrightarrow{\partial} H^{p+1}(Y, j_! \mathbf{Z}/\ell).$$

Given a cohomology class $c \in H^p(X)$ and a smooth point $x \in X$, there exists a Zariski open neighborhood $Y \supset U \ni x$ such that $0 = \partial(c)\big|_U \in H^{p+1}(U, j_! \mathbf{Z}/\ell)$. As an exercise, the reader might work out how this is equivalent to the conjecture for $F = k(X)$. Another exercise is to formulate a version of coniveau filtration as described on page 52 for the group $H^{p+1}(Y, j_! \mathbf{Z}/\ell)$ in such a way that the image of ∂ lies in F^1.

The whole picture of motivic cohomology with finite coefficients is now quite beautiful (Voevodsky et al. 2000). Let X be a smooth, quasi-projective scheme over an algebraically closed field k, and let $m \geq 2$ be relatively prime to the characteristic. Let $r \geq \dim X$. Then

$$H_{\mathrm{M}}^{2r-n}(X, \mathbf{Z}/m\mathbf{Z}(r)) \cong H_{\mathrm{et}}^{2r-n}(X, \mathbf{Z}/m\mathbf{Z}(r)).$$

Said another way, the cycle complexes $\mathcal{Z}^r(X, \bullet)[-2r]$, in equation (0.9), compute the ℓ-adic étale cohomology for all ℓ prime to the characteristic, assuming $r \geq \dim X$. The situation should be compared with that for abelian varieties A where one has Tate modules $T_\ell(A)$ for all ℓ and these calculate $H_1(A, \mathbf{Z}_\ell)$ for ℓ prime to the characteristic.

The subject of torsion in the Chow group seems to be important from many points of view. I include in the bibliography a couple of relevant papers (Soulé and Voisin 2005; Totaro 1997).

Lecture 6: Infinitesimal methods in motivic cohomology

The infinitesimal methods developed here were used also in my work on de Rham–Witt cohomology (Bloch 1977).

It is fair to say that we still do not have an adequate notion of motivic *cohomology*. That is, we do not have contravariant cohomology functors defined on singular schemes (e.g. on non-reduced schemes) with appropriate properties. The notion of $\mathbf{A}^1$-homotopy invariance which is essential in Voevodsky's work is not what one wants. For example, if A is a non-reduced ring, then typically

$$H_{\mathrm{M}}^1(\operatorname{Spec} A, \mathbf{Z}(1)) = A^\times \neq A[t]^\times = H_{\mathrm{M}}^1(\operatorname{Spec} A[t], \mathbf{Z}(1)).$$

Curiously, the K-cohomology groups $H_{\mathrm{Zar}}^p(X, \mathcal{K}_q)$ discussed in Lecture 4 do have the correct functoriality properties, and in this lecture we examine what can be learned from the infinitesimal structure of these groups.

An important step has been the work of Goodwillie (1986) computing the K-theory of nilpotent ideals in characteristic zero in terms of cyclic homology. To understand what motivic cohomology of an infinitesimal thickening might mean, the reader could consult the two rather experimental papers Bloch and

Esnault (1996, 2003). More definitive results have been obtained in Krishna
and Levine (2008), Park (2007), and Rülling (2007).

Assuming one has a good definition of motivic cohomology, what should
the "tangent space"

$$TH^p_M(X, \mathbf{Z}(q)) := \mathrm{Ker}\left(H^p(X \times \mathrm{Spec}\, k[\varepsilon], \mathbf{Z}(q)) \to H^p(X, \mathbf{Z}(q))\right)$$

mean? (Here $\varepsilon^2 = 0$ and the map sends $\varepsilon \mapsto 0$.) To begin with, one should
probably not think of TH_M as a tangent space in the usual sense. It can be
non-trivial in situations where the motivic cohomology itself is rigid, for example for $H^3_M(k, \mathbf{Z}(2))$ with k a number field. Better, perhaps, to think of a
non-semistable moduli functor where jumps can occur at the boundary. For example, consider the Picard scheme of $(\mathbf{P}^1, \{a, b\})$, that is isomorphism classes
of line bundles on $\mathbf{P}^1$ with trivializations at a, b. The degree-zero part is constant $\mathbf{G}_m$ for $a \neq b$, but the limit as $a \to b$ is $\mathbf{G}_a$ given by degree-zero line
bundles on $\mathbf{P}^1$ with double order trivialization at $a = b$.

One important area of open questions about these groups concerns regulators and relations with Euclidean scissors congruence groups. The regulator
for usual motivic cohomology is closely related to volumes in hyperbolic space
(Goncharov 1999), and it seems likely that there is a similar relation between
TH_M and Euclidean volumes. Intuitively, this is another one of those limiting
phenomena where the radius of the hyperbolic disk is allowed to go to infinity
and lengths are scaled so in the limit one gets Euclidean geometry. It would be
nice to have a rigorous description of how this works.

Lecture 7: Diophantine questions

At the time of these lectures, I had expected that the Chow group of zero-cycles
on a rational surface would relate in some way to the zeta function of the surface. As far as I can tell, that does not happen, and I have not thought further
in this direction. The reader who wants to work on diophantine questions regarding zero-cycles and Chow groups should consult Colliot-Thélène (1995),
Esnault (2003), and the references given in these papers.

Lectures 8 and 9: Regulators and values of L-functions

The whole subject of motivic cohomology, regulators, and values of L-functions
remains to a large extent conjectural, but we now understand much better what
should be true (Rapoport et al. 1988; Soulé 1986; Bloch and Kato 1990). For

constructions of the regulator, the reader can consult Goncharov (2005) and Bloch (1986b). Concerning the basic conjecture, I am especially attracted to the formulation given by Fontaine and Perrin-Riou (Fontaine 1992; Fontaine and Perrin-Riou 1994). To understand their idea, let X be a smooth, projective variety of dimension d over $\mathbf{Q}$. (In what follows I gloss over many intractable conjectures.) Consider a motive $M = h^p(X)(q)$, where $h^p(X)$ is a Chow motive as in (0.11), and write $M^*(1) := h^{2d-p}(X)(d-q+1)$. Write $M_B = H^p_B(X, \mathbf{Q}(q))$ for the Betti cohomology of $X(\mathbf{C})$. Let $M_B^+ \subset M_B$ be the subspace invariant under the action of conjugation. Let $t_M := H^p_{DR}(X, \mathbf{Q}(q))/F^0$, where H^*_{DR} is de Rham cohomology. There is a natural map $\alpha \colon M_B^+ \otimes \mathbf{R} \to t_M \otimes \mathbf{R}$, and (assuming certain conjectures) Fontaine and Perrin-Riou construct an exact sequence of motivic cohomology

(0.15)
$$0 \to \mathrm{Hom}_{DM}(\mathbf{Q}(0), M) \otimes \mathbf{R} \to \mathrm{Ker}\,\alpha \to (\mathrm{Ext}^1_{DM,f}(\mathbf{Q}(0), M^*(1)) \otimes \mathbf{R})^*$$
$$\to \mathrm{Ext}^1_{DM,f}(\mathbf{Q}(0), M) \otimes \mathbf{R} \to \mathrm{Coker}\,\alpha \to (\mathrm{Hom}_{DM}(\mathbf{Q}, M^*(1)) \otimes \mathbf{R})^* \to 0.$$

Here Hom and Ext are taken in the triangulated category of Voevodsky motives over $\mathbf{Q}$. The subscript f refers to behavior at finite primes. As a consequence of (0.15) one gets a trivialization over $\mathbf{R}$ of the tensor product of determinant lines

(0.16) $\det(R\,\mathrm{Hom}_{DM,f}(\mathbf{Q}(0), M))_{\mathbf{R}}$
$$\otimes \det(R\,\mathrm{Hom}_{DM,f}(\mathbf{Q}(0), M^*(1)))_{\mathbf{R}} \otimes \det(\alpha)^{-1} \cong \mathbf{R}.$$

The various determinants have $\mathbf{Q}$-structures (though α, itself, does not), so one may examine in (0.16) the ratio of the real trivialization and the rational structure. In fact, using Galois and ℓ-adic cohomology, the authors actually get a $\mathbf{Z}$-structure on the left. They show that the integral conjecture in Bloch and Kato (1990) is equivalent to this ratio being given by $L^*(M, 0)$, the first non-vanishing term in the Taylor expansion of $L(M, s)$ where $L(M, s)$ is the Hasse–Weil L-function associated to M.

Well, okay, there is a lot here we do not understand, but my thought is that one might redo (0.15) working directly with the cycle complexes $\mathcal{Z}^q(X, \bullet)[-2q]$ and $\mathcal{Z}^{d+1-q}(X, \bullet)[-2(d+1-q)]$ which one should think of as concrete realizations of motivic cohomology $R\Gamma_M(X, \mathbf{Z}(q))$ and $R\Gamma_M(X, \mathbf{Z}(d+1-q))$. The resulting determinant metrics could perhaps be deduced (or at least interpreted) directly from the intersection–projection map (well defined in the derived category)

$$\mathcal{Z}^q(X, \bullet)[-2q] \overset{\mathbf{L}}{\otimes} \mathcal{Z}^{d+1-q}(X, \bullet)[-2(d+1-q)] \to \mathcal{Z}^1(\mathrm{Spec}\,\mathbf{Q}, \bullet)[-2d].$$

Hanamura suggested this approach to understanding heights and biextensions. Of course, as it stands it is purely algebraic. It will be necessary to take the cone over the regulator map in some fashion. The result should be some kind of metric or related structure on the determinant of the cycle complex. This would fit well with a conjecture of Soulé (1984), which says in this context that for X regular and proper of dimension d over Spec $\mathbf{Z}$, the Euler characteristic of $\mathcal{Z}^p(X, \bullet)[-2p]$ should be defined and we should have

$$\chi(\mathcal{Z}^p(X, \bullet)[-2p]) = \sum (-1)^i \dim_{\mathbf{Q}} H_{\mathrm{M}}^i(X, \mathbf{Q}(p)) = -\mathrm{ord}_{s=d-p}\zeta_X(s),$$

the negative of the order of zero or pole at $s = d - p$ of the zeta function of X.

Coda: Motives in physics

The subjects of algebraic cycles and motives have enjoyed a tremendous theoretical development over the past 30 years. At the risk of scandalizing the reader, I would say it is high time we start looking for applications.

Dirk Kreimer has been teaching me about Feynman amplitudes and perturbative calculations in quantum field theory. These are periods that arise, for example, in computations of scattering amplitudes. They have a strong tendency to be multi-zeta numbers (Bloch et al. 2006; Broadhurst and Kreimer 1997; but cf. Belkale and Brosnan 2003). The periods in question are associated to graphs. Essentially, the Kirchhoff polynomial (Bloch et al. 2006) of a graph Γ defines a hypersurface X_Γ in projective space, and the Feynman amplitude is a period of this hypersurface relative to a reference symplex. If indeed these are multi-zeta numbers it should be the case that the cohomology of X_Γ has a big Tate piece. One knows if X_Γ were smooth, then $H_{\mathrm{Betti}}^n(X_\Gamma, \mathbf{Q})$ would have pure weight n, so any Tate class, i.e. any map of Hodge structures $\mathbf{Z}(-p) \to H_{\mathrm{Betti}}^n(X_\Gamma, \mathbf{Q})$ would necessarily be a Hodge class, i.e. $n = 2p$. The Hodge conjecture would say that such a class comes from an algebraic cycle, i.e. a class in $H_{\mathrm{M}}^{2p}(X_\Gamma, \mathbf{Z}(p))$ via the realization map from motivic cohomology to Betti cohomology. Unfortunately (or perhaps fortunately), the X_Γ are highly singular, so all one can say are that the weights of H^n are $\leq n$. There are all kinds of possibilities for interesting cohomology classes coming via realization from motivic cohomology. For example, the "wheel with n spokes graph" $WS(n)$ gives rise to a hypersurface $X_{WS(n)}$ of dimension $2n - 2$. The primitive cohomology in the middle dimension for this graph is computed (Bloch et al. 2006) to be $\mathbf{Q}(-2)$ (independent of n). An appropriate generalization of the Hodge conjecture would suggest a class in $H_{\mathrm{M}}^{2n-2}(X_{WS(n)}, \mathbf{Q}(2))$. The problem of computing such motivic cohomology groups can be attacked via the combinatorics of the graph, but what has so far proved more powerful is to use classical algebro-geometric techniques to study the geometry of rank stratifications associated to a homomorphism of vector bundles $u: E \to F$ over projective space.

Ideally, knowledge of motives should provide a strong organizing force to study complex physical phenomena. Even simple motivic invariants like weight and Hodge level should help physicists understand the periods arising in their computations. More sophisticated methods involving monodromy and limiting mixed motives may give information about Landau singularities and unitarity of the S-matrix.

References for preface

[1] Beĭlinson, A. A. 1985. Higher regulators and values of L-functions. *J. Soviet Math.*, **30**, 2036–2070.

[2] Beĭlinson, A. A., A. B. Goncharov, V. V. Schechtman, and A. N. Varchenko. 1990. Aomoto dilogarithms, mixed Hodge structures and motivic cohomology of pairs of triangles on the plane. Pp. 135–172 in *The Grothendieck Festschrift, Vol. I*. Progr. Math., vol. 86. Boston: Birkhäuser Boston.

[3] Belkale, Prakash, and Patrick Brosnan. 2003. Matroids, motives, and a conjecture of Kontsevich. *Duke Math. J.*, **116**(1), 147–188.

[4] Bloch, Spencer. 1977. Algebraic K-theory and crystalline cohomology. *Inst. Hautes Études Sci. Publ. Math.*, 187–268 (1978).

[5] Bloch, S. 1980. Algebraic K-theory and zeta functions of elliptic curves. Pp. 511–515 in *Proceedings of the International Congress of Mathematicians (Helsinki, 1978)*. Helsinki: Acad. Sci. Fennica.

[6] Bloch, S. 1994. The moving lemma for higher Chow groups. *J. Algebraic Geom.*, **3**(3), 537–568.

[7] Bloch, Spencer. 1986a. Algebraic cycles and higher K-theory. *Adv. in Math.*, **61**(3), 267–304.

[8] Bloch, Spencer. 1986b. Algebraic cycles and the Beĭlinson conjectures. Pp. 65–79 in *The Lefschetz Centennial Conference, Part I (Mexico City, 1984)*. Contemp. Math., vol. 58. Providence, R.I.: American Mathematical Society.

[9] Bloch, Spencer, and Hélène Esnault. 1996. The coniveau filtration and non-divisibility for algebraic cycles. *Math. Ann.*, **304**(2), 303–314.

[10] Bloch, Spencer J. 2000. *Higher Regulators, Algebraic K-Theory, and*

Zeta Functions of Elliptic Curves. CRM Monograph Series, vol. 11. Providence, R.I.: American Mathematical Society.

[11] Bloch, Spencer, and Hélène Esnault. 2003. An additive version of higher Chow groups. *Ann. Sci. École Norm. Sup. (4)*, **36**(3), 463–477.

[12] Bloch, Spencer, Hélène Esnault, and Dirk Kreimer. 2006. On motives associated to graph polynomials. *Comm. Math. Phys.*, **267**(1), 181–225.

[13] Bloch, Spencer, and Kazuya Kato. 1990. L-functions and Tamagawa numbers of motives. Pp. 333–400 in *The Grothendieck Festschrift, Vol. I*. Progr. Math., vol. 86. Boston: Birkhäuser Boston.

[14] Bloch, Spencer, and Igor Kříž. 1994. Mixed Tate motives. *Ann. of Math. (2)*, **140**(3), 557–605.

[15] Borel, Armand. 1977. Cohomologie de SL_n et valeurs de fonctions zeta aux points entiers. *Ann. Scuola Norm. Sup. Pisa Cl. Sci. (4)*, **4**(4), 613–636.

[16] Broadhurst, D. J., and D. Kreimer. 1997. Association of multiple zeta values with positive knots via Feynman diagrams up to 9 loops. *Phys. Lett. B*, **393**(3–4), 403–412.

[17] Carlson, James A. 1987. The geometry of the extension class of a mixed Hodge structure. Pp. 199–222 in *Algebraic Geometry, Bowdoin, 1985 (Brunswick, Maine, 1985)*. Proc. Sympos. Pure Math., vol. 46. Providence, R.I.: American Mathematical Society.

[18] Carlson, James A., and Richard M. Hain. 1989. Extensions of variations of mixed Hodge structure. *Astérisque*, **9**, 39–65. Actes du Colloque de Théorie de Hodge (Luminy, 1987).

[19] Cattani, Eduardo, Aroldo Kaplan, and Wilfried Schmid. 1986. Degeneration of Hodge structures. *Ann. of Math. (2)*, **123**(3), 457–535.

[20] Clemens, Herbert. 1983. Homological equivalence, modulo algebraic equivalence, is not finitely generated. *Inst. Hautes Études Sci. Publ. Math.*, 19–38 (1984).

[21] Colliot-Thélène, Jean-Louis. 1995. L'arithmétique des zéro-cycles (exposé aux Journées arithmétiques de Bordeaux, septembre 93). *Journal de théorie des nombres de Bordeaux*, 51–73.

[22] Deligne, Pierre, and Alexander B. Goncharov. 2005. Groupes fondamentaux motiviques de Tate mixte. *Ann. Sci. École Norm. Sup. (4)*, **38**(1), 1–56.

[23] Esnault, Hélène. 2003. Varieties over a finite field with trivial Chow group of 0-cycles have a rational point. *Invent. Math.*, **151**(1), 187–191.

[24] Fontaine, Jean-Marc. 1992. Valeurs spéciales des fonctions L des motifs. *Astérisque*, exp. no. 751, 4, 205–249. Séminaire Bourbaki, vol. 1991/92.

[25] Fontaine, Jean-Marc, and Bernadette Perrin-Riou. 1994. Autour des conjectures de Bloch et Kato: cohomologie galoisienne et valeurs de fonctions *L*. Pp. 599–706 in *Motives (Seattle, 1991)*. Proc. Sympos. Pure Math., vol. 55. Providence, R.I.: American Mathematical Society.

[26] Friedlander, Eric M. 2005. Motivic complexes of Suslin and Voevodsky. Pp. 1081–1104 in *Handbook of K-Theory. Vol. 1, 2*. Berlin: Springer.

[27] Goncharov, Alexander. 1999. Volumes of hyperbolic manifolds and mixed Tate motives. *J. Amer. Math. Soc.*, **12**(2), 569–618.

[28] Goncharov, Alexander B. 2005. Regulators. Pp. 295–349 in *Handbook of K-Theory. Vol. 1, 2*. Berlin: Springer.

[29] Goodwillie, Thomas G. 1986. Relative algebraic K-theory and cyclic homology. *Ann. of Math. (2)*, **124**(2), 347–402.

[30] Hanamura, Masaki. 1995. Mixed motives and algebraic cycles. I. *Math. Res. Lett.*, **2**(6), 811–821.

[31] Hanamura, Masaki. 1999. Mixed motives and algebraic cycles. III. *Math. Res. Lett.*, **6**(1), 61–82.

[32] Hanamura, Masaki. 2000. Homological and cohomological motives of algebraic varieties. *Invent. Math.*, **142**(2), 319–349.

[33] Jannsen, Uwe. 1994. Motivic sheaves and filtrations on Chow groups. Pp. 245–302 in *Motives (Seattle, 1991)*. Proc. Sympos. Pure Math., vol. 55. Providence, R.I.: American Mathematical Society.

[34] Kahn, Bruno. 1996. Applications of weight-two motivic cohomology. *Doc. Math.*, **1**(17), 395–416.

[35] Krishna, Amalendu, and Marc Levine. 2008. Additive higher Chow groups of schemes. *J. Reine Angew. Math.*, **619**, 75–140.

[36] Levine, Marc. 1998. *Mixed Motives*. Mathematical Surveys and Monographs, vol. 57. Providence, R.I.: American Mathematical Society.

[37] Mazza, Carlo, Vladimir Voevodsky, and Charles Weibel. 2006. *Lecture Notes on Motivic Cohomology*. Clay Mathematics Monographs, vol. 2. Providence, R.I.: American Mathematical Society.

[38] Murre, J. P. 1993a. On a conjectural filtration on the Chow groups of an algebraic variety. I. The general conjectures and some examples. *Indag. Math. (N.S.)*, **4**(2), 177–188.

[39] Murre, J. P. 1993b. On a conjectural filtration on the Chow groups of an algebraic variety. II. Verification of the conjectures for threefolds which are the product on a surface and a curve. *Indag. Math. (N.S.)*, **4**(2), 189–201.

[40] Nori, Madhav V. 1993. Algebraic cycles and Hodge-theoretic connectivity. *Invent. Math.*, **111**(2), 349–373.

[41] Park, Jinhyun. 2007. Algebraic cycles and additive dilogarithm. *Int. Math. Res. Not. IMRN*, Art. ID rnm067, 19.

[42] Rapoport, M., N. Schappacher, and P. Schneider (eds). 1988. *Beilinson's Conjectures on Special Values of L-Functions*. Perspectives in Mathematics, vol. 4. Boston: Academic Press.

[43] Rülling, Kay. 2007. The generalized de Rham-Witt complex over a field is a complex of zero-cycles. *J. Algebraic Geom.*, **16**(1), 109–169.

[44] Soulé, Christophe. 1984. K-théorie et zéros aux points entiers de fonctions zêta. Pp. 437–445 in *Proceedings of the International Congress of Mathematicians, Vol. 1, 2 (Warsaw, 1983)*. Warsaw: PWN.

[45] Soulé, Christophe. 1986. Régulateurs. *Astérisque*, 237–253. Seminar Bourbaki, vol. 1984/85.

[46] Soulé, C., and C. Voisin. 2005. Torsion cohomology classes and algebraic cycles on complex projective manifolds. *Adv. Math.*, **198**(1), 107–127.

[47] Totaro, Burt. 1997. Torsion algebraic cycles and complex cobordism. *J. Amer. Math. Soc.*, **10**(2), 467–493.

[48] Totaro, Burt. 1999. The Chow ring of a classifying space. Pp. 249–281 in *Algebraic K-Theory (Seattle, 1997)*. Proc. Sympos. Pure Math., vol. 67. Providence, R.I.: American Mathematical Society.

[49] Voevodsky, Vladimir, Andrei Suslin, and Eric Friedlander. 2000. *Cycles, Transfers, and Motivic Homology Theories*. Annals of Mathematics Studies, vol. 143. Princeton, N.J.: Princeton University Press.

[50] Voisin, Claire. 1994. Transcendental methods in the study of algebraic cycles. Pp. 153–222 in *Algebraic Cycles and Hodge Theory (Torino, 1993)*. Lecture Notes in Math., no. 1594. Berlin: Springer.

[51] Voisin, Claire. 2004. Remarks on filtrations on Chow groups and the Bloch conjecture. *Ann. Mat. Pura Appl. (4)*, **183**(3), 421–438.

LECTURES ON ALGEBRAIC CYCLES

0

Introduction

The notes which follow are based on lectures given at Duke University in April, 1979. They are the fruit of ten years reflection on *algebraic cycles*; that is formal linear combinations $\sum n_i [V_i]$ of subvarieties V_i of a fixed smooth and projective variety X, with integer coefficients n_i.

Classically, X was an algebraic curve (Riemann surface) and the V_i were points p_i. In this context, cycles were a principal object of study for nineteenth-century complex function theory. Two (at least) of their results merit immortality: the theorems of Riemann–Roch and Abel–Jacobi.

The reader will recall that to a meromorphic function f on a smooth variety or complex manifold one can associate a *divisor* (algebraic cycle of codimension 1) (f) by taking the sum of its components of zeros minus its components of poles, all counted with suitable multiplicity. An algebraic cycle $\sum n_i [V_i]$ is said to be *effective* if all the $n_i \geq 0$. The Riemann–Roch theorem computes the dimension $\ell(D)$ of the vector space of functions f such that $(f) + D$ is effective for a given divisor D. It has been generalized quite considerably in recent years, but its central role in the study of divisors does not seem to carry over to cycles of codimension greater than 1. (For example, one can use the Riemann–Roch theorem to prove rationality of the zeta function of an algebraic curve over a finite field [9], but the corresponding theorem for varieties of dimension > 1 lies deeper.)

The other great pillar of function theory on Riemann surfaces, the Abel–Jacobi theorem, tells when a given divisor $D = \sum n_i (p_i)$ is the divisor of a function. Clearly a necessary condition is that the degree $\sum n_i$ must be zero, so we may assume this and write our cycle $D = \sum n_i ((p_i) - (p_0))$ for some base point p_0. We denote by $\Gamma(X, \Omega_X^1)$ the space of all global holomorphic differential 1-forms on X (where $\omega \in \Gamma(X, \Omega_X^1)$ can be written locally as $f(z)\,dz$ for z and $f(z)$ holomorphic). A necessary and sufficient condition for D to equal

(f) is that

$$\sum n_i \int_{p_0}^{p_i} \omega = \int_\gamma \omega$$

for some (topological) 1-cycle $\gamma \in H_1(X, \mathbf{Z})$ and all $\omega \in \Gamma(X, \Omega_X^1)$. Even more is true; writing $\Gamma(X, \Omega_X^1)^* = \mathrm{Hom}_{\mathbf{C}}(\Gamma(X, \Omega_X^1), \mathbf{C})$ for the *periods*, the map

$$A_0(X) := \frac{\text{divisors of degree 0}}{\text{divisors of functions}} \to \Gamma(X, \Omega_X^1)^* / H_1(X, \mathbf{Z}) := J(X)$$

is an isomorphism from $A_0(X)$ to the *abelian variety* $J(X)$, where $J(X)$ is the *jacobian* of X.

To extend these results, we may define for X any smooth projective variety over a field and $r \geq 0$ an integer

$$z^r(X) = \text{free abelian group generated by irreducible}$$
$$\text{subvarieties of } X \text{ of codimension } r.$$

Given $A = \sum m_i A_i \in z^r(X)$ and $B = \sum n_j B_j \in z^s(X)$ such that every component of $A_i \cap B_j$ has codimension $r + s$ for all i and j, there is defined a product cycle $A \cdot B \in z^{r+s}(X)$ obtained by summing over all components of $A_i \cap B_j$ with suitable multiplicities. If $f : X \to Y$ is a proper map, where $\dim X = d$, $\dim Y = e$, one has $f_* : z^r(X) \to z^{r+e-d}(Y)$ defined on a single irreducible codimension-r subvariety $V \subset X$ by $f_*(V) = [k(V) : k(f(V))] \cdot f(V)$. Here $[k(V) : k(f(V))]$ is the degree of the extension of function fields. By convention this degree is zero if the extension is not finite.

Suppose now $\Gamma \in z^r(X \times \mathbf{P}^1)$ and no component of Γ contains either $X \times \{0\}$ or $X \times \{\infty\}$. The cycle

$$\mathrm{pr}_{1*}(\Gamma \cdot (X \times ((0) - (\infty)))) \in z^r(X), \qquad \mathrm{pr}_1 : X \times \mathbf{P}^1 \to X,$$

is then defined. By definition, $z_{\mathrm{rat}}^r(X) \subset z^r(X)$ is the subgroup of cycles of this form. Replacing $\mathbf{P}^1$ by an arbitrary (variable) smooth connected curve C, and $0, \infty$ by any two points $a, b \in C$ one obtains a group $z_{\mathrm{alg}}^r(X)$ with

$$z_{\mathrm{rat}}^r(X) \subset z_{\mathrm{alg}}^r(X) \subset z^r(X).$$

We write

$$\mathrm{CH}^r(X) = z^r(X)/z_{\mathrm{rat}}^r(X) \qquad \text{(the *Chow group*)},$$

$$A^r(X) = z_{\mathrm{alg}}^r(X)/z_{\mathrm{rat}}^r(X),$$

$$\mathrm{CH}_r(X) = \mathrm{CH}^{d-r}(X),$$

$$A_r(X) = A^{d-r}(X), \qquad d = \dim X.$$

Cycles in $z_{rat}^r(X)$ (resp. $z_{alg}^r(X)$) are said to be *rationally* (resp. *algebraically*) *equivalent to zero* (written $x \underset{rat}{\sim} 0$ or $x \underset{alg}{\sim} 0$, or if no ambiguity is possible just $x \sim 0$). As an exercise, the reader might prove that when the ground field k is algebraically closed, the association $\sum n_i (p_i) \mapsto \sum n_i$ defines an isomorphism $CH_0(X)/A_0(X) \cong \mathbf{Z}$. A bit more difficult is to show

$$CH^1(X) \cong \frac{\text{divisors}}{\text{divisors of functions}} = \text{divisor class group of } X \cong \text{Pic}(X),$$

where $\text{Pic}(X)$ is the group of isomorphism classes of line bundles on X. Still harder is to prove that when X has a rational point, $A^1(X)$ is isomorphic to the group of k-points of the *Picard variety* of X.

The purpose of these notes is to study algebraic cycles, *particularly those of codimension* > 1 as it is here that really new and unexpected phenomena occur. In keeping with the author's philosophy that good mathematics "opens out" and involves several branches of the subject, we will consider geometric, algebraic and arithmetic problems.

The first three lectures are geometric in content, studying various aspects of the Abel–Jacobi construction in codimensions > 1. We work with varieties X over the complex numbers, and define (following Griffiths, Lieberman, and Weil) compact complex tori (*intermediate jacobians*) $J^r(X)$ and *Abel–Jacobi maps*

$$\Theta: A^r(X) \to J^r(X).$$

When $r = d = \dim X$, $J^r(X) = J_0(X) = \Gamma(X, \Omega_X^1)^*/H_1(X, \mathbf{Z})$ is called the Albanese variety and the Abel–Jacobi map is defined precisely as with curves. The map $\Theta: A_0(X) \to J_0(X)$ is surjective, and in the first lecture we study its kernel $T(X)$ when $\dim X = 2$. We sketch an argument of Mumford that $T(X) \neq 0$ (in fact $T(X)$ is enormous) when $P_g(X) \neq 0$, i.e. when X has a non-zero global holomorphic 2-form. We give several examples motivating the conjecture that $\Gamma(X, \Omega_X^2)$ actually controls the structure of $T(X)$ and that in particular $T(X) = 0 \iff \Gamma(X, \Omega_X^2) = 0$. This conjecture can be formulated in various ways. One vague but exciting possibility is that groups like $T(X)$ provide a geometric interpretation of the category of *polarized Hodge structures of weight two*, in much the same way that abelian varieties do for weight 1.

The second and third lectures consider curves on threefolds. In Lecture 2 we focus on quartic threefolds, i.e. smooth hypersurfaces X of degree 4 in $\mathbf{P}^4$, and verify in that case Θ is an isomorphism. Lecture 3 considers *relative* algebraic 1-cycles on $X = C \times \mathbf{P}^1 \times \mathbf{P}^1$, where C is a curve. We define a *relative intermediate jacobian* which turns out to be a non-compact torus isomorphic to $H^1(C, \mathbf{C}^*)$. The machinery of cycle classes constructed in this lecture is related in Lectures 8 and 9 to special values of Hasse–Weil zeta functions. Of particular importance are the classes in $H^1(C, \mathbf{R})$ defined by factoring out by the maximal compact. We show such classes can be defined for any (not necessarily relative) cycle on $C \times \mathbf{P}^1 \times \mathbf{P}^1$.

Lectures 4 through 6 develop the algebraic side of the theory: the cohomology groups of the K-sheaves $H^p(X, \mathcal{K}_q)$, the Gersten–Quillen resolution, and analogues for singular and étale cohomology theories. Of particular interest are the cohomological formulae

$$CH^p(X) \cong H^p(X, \mathcal{K}_p)$$

and (for X defined over $\mathbf{C}$)

$$CH^p(X)/A^p(X) \cong H^p(X, \mathcal{H}^p),$$

where $\mathcal{H}^p$ denotes the sheaf for the Zariski topology associated to the presheaf $U \subset X \to H^p(U, \mathbf{Z})$ (singular cohomology). These techniques are used to prove a theorem of Roitman that $A_0(X)_{\text{tors}}$, the torsion subgroup of the zero-cycles, maps isomorphically to the torsion subgroup of the Albanese variety. The heart of the proof is a result about the multiplicative structure of the Galois cohomology *ring* of a function field F of transcendence degree d over an algebraically closed field. We prove for ℓ prime to char F that the cup product map

$$\underbrace{H^1(F, \mathbf{Z}/\ell\mathbf{Z}) \otimes \cdots \otimes H^1(F, \mathbf{Z}/\ell\mathbf{Z})}_{d \text{ times}} \to H^d(F, \mathbf{Z}/\ell\mathbf{Z})$$

is surjective. It would be of great interest both algebraically and geometrically to know if the whole cohomology ring $H^*(F, \mathbf{Z}/\ell\mathbf{Z})$ were generated by H^1.

The last three lectures are devoted to arithmetic questions. In Lecture 7 we consider $A_0(X)$, where X is a surface over a local or global field k. We assume the base extension of X to the algebraic closure of k is a rational surface. Using a technique of Manin involving the *Brauer group* we show by example that in general $A_0(X) \neq 0$. We prove that $A_0(X)$ is finite when X has a pencil of genus-zero curves (*conic bundle surface*). The key idea in the proof is a sort of generalization of the *Eichler norm theorem*, describing the image of the reduced norm map Nrd: $A^* \to k(t)^*$ when A is a quaternion algebra defined over a rational field in one variable over k.

Finally, Lectures 8 and 9 take up, from a number-theoretic point of view, the work of Lecture 3 on relative intermediate jacobians for curves on $C \times \mathbf{P}^1 \times \mathbf{P}^1$. We consider the case $C = E =$ elliptic curve and compute explicitly the class in $H^1(E, \mathbf{R})$ associated to the curve

$$\gamma_{f,g} = \{(x, f(x), g(x)) \mid x \in E\}$$

for f, g rational functions on E. When the zeros and poles of f are points of finite order, we show how to associate to f and g an element in $\Gamma(E, \mathcal{K}_2)$ and how to associate to an element in $\Gamma(E, \mathcal{K}_2)$ a relative algebraic 1-cycle on $E \times \mathbf{P}^1 \times \mathbf{P}^1$. When E has complex multiplication by the ring of integers in an imaginary quadratic field of class number one, we construct an element $U \in \Gamma(E_\mathbf{Q}, \mathcal{K}_2)$ such that the image under the Abel–Jacobi map into $H^1(E, \mathbf{R}) \cong \mathbf{C}$ of the associated relative algebraic cycle multiplied by a certain simple constant (involving the conductor of the curve and Gauss sum) is the value of the Hasse–Weil zeta function of E at $s = 2$. Conjecturally for E defined over a number field k, the rank of $\Gamma(E_k, \mathcal{K}_2)$ equals the order of the zero of the Hasse–Weil zeta function of E at $s = 0$.

I want to thank Duke University for financial support, and my auditors at Duke for their enthusiasm and tenacity in attending eight lectures in ten days. I also want to acknowledge that much of the function theory of the dilogarithm which underlines the calculations in the last two lectures was worked out in collaboration with David Wigner, and that I never would have gotten the damned constants in (9.12) correct without help from Dick Gross. (I may, indeed, not have gotten them correct even with help.)

Finally, a pedagogical note. When an idea is already well documented in the literature and extensive detail would carry us away from the focus of these notes on algebraic cycles, I have been very sketchy. For example, Quillen's work on the foundations of K-theory are magnificently presented in his own paper [10]. I only hope the brief outline given here will motivate the reader to turn to that source. The reader may also find the rapid treatment of the Mumford argument showing $P_g \neq 0 \Rightarrow T(X) \neq 0$ in Lecture 1 to be unsatisfactory. To remedy this, another demonstration rather different in spirit from Mumford's is included as an appendix. Historically, the negative force of this result led us all to conclude that, except in certain obvious cases such as ruled surfaces, the structure of zero-cycles on a surface was total chaos. I wanted to devote time to various examples showing that this is not the case. In retrospect, I have certainly "left undone those things which I ought to have done" (e.g. Griffiths' proof that homological equivalence $\neq$ algebraic equivalence, and Tate's proof of the Tate conjecture for abelian varieties). I hope the reader will spare me.

References for Lecture 0

A list of references appears at the end of each lecture. What follows are references for foundational questions about cycles, etc.

[1] C. Chevalley et al., Anneaux de Chow et applications, *Séminaire C. Chevalley*, 2e année, Sécr. Math. Paris (1958).

[2] W. L. Chow, On equivalence classes of cycles in an algebraic variety, *Ann. of Math. (2)*, **64**, 450–479 (1956).

[3] A. Grothendieck et al., *Théorie des intersections et théorème de Riemann–Roch* (SGA 6), Lecture Notes in Math., no. 225, Springer, Berlin (1971).

[4] W. Fulton, Rational equivalence on singular varieties, *Inst. Hautes Études Sci. Publ. Math.*, no. 45 (1975), 147–167.

[5] W. Fulton and R. MacPherson, Intersecting cycles on an algebraic variety, Aarhus Universitet Preprint Series, no. 14 (1976). [Pp. 179–197 in *Real and complex singularities (Proc. Ninth Nordic Summer School/NAVF Sympos. Math., Oslo, 1976)*, Sijthoff and Noordhoff, Alphen aan den Rijn (1977).]

[6] P. Samuel, Rational equivalence of arbitrary cycles, *Amer. J. Math.*, **78** (1956), 383–400.

[7] J.-P. Serre, *Algèbre locale. Multiplicités*, Lecture Notes in Math., no. 11, Springer, Berlin (1965).

[8] A. Weil, *Foundations of Algebraic Geometry*, A.M.S. Colloquium Publications, vol. 29, American Mathematical Society, Providence, R.I. (1946).

[9] A. Weil, *Courbes Algébriques et Variétés Abéliennes*, Hermann, Paris (1971).

[10] D. Quillen, Higher algebraic K-theory. I, pp. 85–147 in *Algebraic K-Theory I*, Lecture Notes in Math., no. 341, Springer, Berlin (1973).

1

Zero-cycles on surfaces

In this first section I want to consider the question of zero-cycles on an algebraic surface from a purely geometric point of view. I will consider a number of explicit examples, and give a heuristic description of a result of Mumford [4] that $P_g \neq 0$ implies $A_0(X)$ is "very large". In particular $A_0(X)$ is not an abelian variety in this case. Finally I will discuss some conjectures motivated by these ideas.

For purposes of this lecture, algebraic surface X will mean a smooth projective variety of dimension 2 over the complex numbers (or, if you prefer, a compact complex manifold of dimension 2 admitting a projective embedding). The space of global holomorphic i-forms ($i = 1, 2$) will be written $\Gamma(X, \Omega_X^i)$ (or $H^{i,0}$) and we will write

$$q = \dim \Gamma(X, \Omega_X^1), \quad P_g = \dim \Gamma(X, \Omega_X^2).$$

If γ is a topological 1-chain on X, the integral over γ, $\int_\gamma$, is a well-defined element in the dual $\mathbf{C}$-vector space $\Gamma(X, \Omega_X^1)^*$. If C is a 2-chain,

$$\int_{\partial C} \omega = \int_C d\omega \qquad \text{(Stokes')}$$

so $\int_{\partial C} = 0$ in $\Gamma(X, \Omega_X^1)^*$, since global holomorphic forms are closed. Hence we may map $H_1(X, \mathbf{Z}) \to \Gamma(X, \Omega_X^1)^*$. It follows from the Hodge-theoretic decomposition $H^1(X, \mathbf{C}) = H^{0,1} \oplus H^{1,0}$ that the quotient $\Gamma(X, \Omega_X^1)^*/H_1(X, \mathbf{Z})$ is a compact complex torus, called the Albanese of X and written $\mathrm{Alb}(X)$. This torus admits a polarization satisfying the Riemann bilinear relations and hence is an abelian variety.

Fixing a base point $p_0 \in X$, we define $\phi \colon X \to \mathrm{Alb}(X)$ by $\phi(p) = \int_{p_0}^p$. Note that this is well defined: two paths from p_0 to p differ by an element in $H_1(X, \mathbf{Z})$. Notice, finally, that the above discussion is equally valid for smooth projective varieties X of arbitrary dimension.

Proposition (1.1)

(i) Let $\phi_n\colon \underbrace{X \times \cdots \times X}_{n \text{ times}} \to \mathrm{Alb}(X)$ be defined by $\phi_n(x_1, \ldots, x_n) = \phi(x_1) + \cdots + \phi(x_n)$. Then for $n \gg 0$, ϕ_n is surjective.

(ii) Let $\psi\colon X \to A$ be a map from X to a complex torus, and assume $\psi(p_0) = 0$. Then there exists a unique homomorphism $\theta\colon \mathrm{Alb}(X) \to A$ such that the diagram

$$
\begin{array}{ccc}
X & \xrightarrow{\ \phi\ } & \mathrm{Alb}(X) \\
\psi \downarrow & \swarrow \theta & \\
A & &
\end{array}
$$

commutes.

(iii) The map $z_0(X) \xrightarrow{\phi} \mathrm{Alb}(X)$, where $(x) \mapsto \phi(x)$, factors through the map $\phi\colon \mathrm{CH}_0(X) \to \mathrm{Alb}(X)$. The induced map $\phi\colon A_0(X) \to \mathrm{Alb}(X)$ is surjective, and independent of the choice of base point p_0.

Proof These results are more or less well known. The reader who is unfamiliar with them might try as an exercise to find proofs. (Hint: In (i) consider the question infinitesimally and use the fundamental theorem of calculus to calculate the derivative of $\int_{p_0}^{p} \omega$. For (iii), reduce the question to showing that a map $\mathbf{P}^1 \to$ complex torus is necessarily constant.) $\qquad\square$

Example (1.2) (Bloch et al. [2]) Let E and F be elliptic curves (Riemann surfaces of genus 1). We propose to calculate the Chow group of the quotient surface $X = (F \times E)/\{1, \sigma\}$, where σ is a fixed-point-free involution on $F \times E$ obtained by fixing a point $\eta \in E$ of order 2, $\eta \neq 0$, and taking $\sigma(f, e) = (-f, e + \eta)$. Let $E' = E/\{1, \eta\}$. There is a natural map $\rho : X \to E'$ with all fibres of $\rho \cong F$.

Notice

$$
\Gamma(X, \Omega_X^1) \cong \Gamma(F \times E, \Omega_{F \times E}^1)^{\{1, \sigma\}}
$$
$$
\cong \left[\Gamma(E, \Omega_E^1) \oplus \Gamma(F, \Omega_F^1) \right]^{\{1, \sigma\}}
$$
$$
\cong \Gamma(E, \Omega_E^1) \cong \mathbf{C}
$$

since the automorphism $f \to -f$ acts by -1 on $\Gamma(F, \Omega_F^1)$. We conclude $\mathrm{Alb}(X)$ has dimension 1. Since $\mathrm{Alb}(X) \to E'$ and the fibres of ρ are connected, it follows that $\mathrm{Alb}(X) \cong E'$.

Lemma (1.3) *Let Y be a smooth quasi-projective variety, $n > 0$ an integer. Then $A_n(X)$ is a divisible group.*

Proof By definition $A_n(X) \subset CH_n(X)$ is generated by images under correspondences from jacobians of curves. Since jacobians of curves are divisible groups, the lemma follows. $\square$

Lemma (1.4) *Let Y be a smooth projective variety, and let*

$$T(Y) = \mathrm{Ker}\,(A_0(Y) \to \mathrm{Alb}(Y)).$$

Then $T(Y)$ is divisible.

Proof For any abelian group A, let $_NA \subset A$ be the kernel of multiplication by N. From the divisibility of $A_0(Y)$ one reduces to showing $_NA_0(Y) \twoheadrightarrow {}_N\mathrm{Alb}(Y)$ for any N. If Y is a curve, the Chow group and Albanese both coincide with the jacobian so $_NA_0(Y) \cong {}_N\mathrm{Alb}(Y)$. It will suffice therefore to assume $\dim Y > 1$ and show $_N\mathrm{Alb}(W) \twoheadrightarrow {}_N\mathrm{Alb}(Y)$ for $W \subset Y$ a smooth hyperplane section. As a real torus, $\mathrm{Alb}(Y)$ can be identified with $H_1(Y, \mathbf{R}/\mathbf{Z})$ and $_N\mathrm{Alb}(Y) = H_1(Y, \mathbf{Z}/N\mathbf{Z})$, so the question becomes the surjectivity of $H_1(W, \mathbf{Z}/N\mathbf{Z}) \to H_1(Y, \mathbf{Z}/N\mathbf{Z})$ or the vanishing of $H_1(Y, W; \mathbf{Z}/N\mathbf{Z})$. This group is identified by duality with $H^{2\dim Y-1}(Y - W, \mathbf{Z}/N\mathbf{Z}) = (0)$ because the Stein variety $Y - W$ has cohomological dimension $= \dim Y$. $\square$

We now return to our surface $X = F \times E/\{1, \sigma\}$.

Claim $A_0(X) \cong E' \cong \mathrm{Alb}(X)$. *That is, $T(X) = 0$.*

Proof We work with the diagram

$$
\begin{array}{ccc}
F \times E & \longrightarrow & E \\
\pi \downarrow & & \downarrow \\
X & \longrightarrow & E'.
\end{array}
$$

If $z \in T(X)$ we may write

$$\pi^*z = \sum r_i\,[(q_i, p_i) + (-q_i, p_i + \eta)],$$

where $\sum r_i = 0$ and $\sum 2\,r_i p_i = 0$ in E. By Abel's theorem

$$2\,(q, p) \sim 2\,(q, p + \eta) \qquad \text{on } F \times E,$$
$$(q, p) + (-q, p) \sim 2\,(0, p),$$

so

$$2\,\pi^*z \sim \sum 2\,r_i\,[(q_i, p_i) + (-q_i, p_i)] \sim \sum 4\,r_i\,(0, p_i)$$
$$= [(0) \times E] \cdot \Big[F \times \sum 4\,r_i\,(p_i) \Big]$$
$$\sim 0\,.$$

It follows that $0 \sim 2\pi_*\pi^*z = 4z$, and hence $4T(X) = 0$. Since $T(X)$ is divisible, this implies $T(X) = 0$.   □

Example (1.5) (Inose and Mizukami [3])   Let $Y : T_0^5 + T_1^5 + T_2^5 + T_3^5 = 0$ in $\mathbf{P}^3$, and let $X = Y/(\mathbf{Z}/5\mathbf{Z})$, where we identify $\mathbf{Z}/5\mathbf{Z}$ with the group of fifth roots of 1 and we let a fifth root ω act on Y by $(t_0, t_1, t_2, t_3) \rightarrow (t_0, \omega t_1, \omega^2 t_2, \omega^3 t_3)$. This action is fixed point free, so X is smooth.

Claim $A_0(X) = (0)$.

Proof   Let $\pi: Y \rightarrow X$ be the projection. Since $A_0(X)$ is divisible, it suffices to show $5A_0(X) = \pi_*\pi^*A_0(X) = (0)$. Since π_* is surjective, it suffices to show $\pi^*\pi_*: A_0(Y) \rightarrow A_0(Y)$ is the zero map. We fix a fifth root of 1, $\omega \neq 1$, and let $\mathbf{Z}/5\mathbf{Z}^{\oplus 3}$ act on Y by

$$e_i^* T_j = \begin{cases} T_j & j \neq i, \\ \omega T_i & i = j, \end{cases} \quad i = 1, 2, 3.$$

This gives us a representation $\mathbf{Z}[\mathbf{Z}/5\mathbf{Z}^{\oplus 3}] \twoheadrightarrow R \subset \text{End}(A_0(Y))$, which we denote by $x \mapsto \bar{x}$. One checks immediately

(1.5.1) $\pi^*\pi_* = 1 + \bar{e}_1\bar{e}_2^2\bar{e}_3^3 + (\bar{e}_1\bar{e}_2^2\bar{e}_3^3)^2 + \cdots + (\bar{e}_1\bar{e}_2^2\bar{e}_3^3)^4.$

The reader can verify also that the quotient of Y by any of the cyclic groups

$$\langle e_1 e_2 e_3, \ldots, e_1^5 e_2^5 e_3^5 \rangle \quad \text{or}$$
$$\langle e_i, e_i^2, \ldots, e_i^5 \rangle, \quad i = 1, 2, 3 \quad \text{or}$$
$$\langle e_i e_j, e_i^2 e_j^2, \ldots, e_i^5 e_j^5 \rangle, \quad i \neq j$$

is rational, leading to

$$0 = \sum_{n=0}^{4} \bar{e}_1^n \bar{e}_2^n \bar{e}_3^n,$$

(1.5.2) $\displaystyle\sum_{n=0}^{4} e_i^{-n} = 0 = \sum_{n=0}^{4} \bar{e}_i^n \bar{e}_j^n$ in $R \subset \text{End}(A_0(Y)).$

Since $A_0(Y)$ is divisible, it will suffice to show $\pi^*\pi_*$ (1.5.1) is trivial in $R \otimes \mathbf{Q}$. Writing $F = \mathbf{Q}(\omega)$, the left-hand identities in (1.5.2) show that $R \otimes \mathbf{Q}$ is a quotient of the semi-simple ring $F \otimes_{\mathbf{Q}} F \otimes_{\mathbf{Q}} F$, so it will suffice to show that $\pi^*\pi_*$ goes to zero under any homomorphism $R \otimes \mathbf{Q} \rightarrow \bar{\mathbf{Q}} = $ algebraic closure of $\mathbf{Q}$. A homomorphism $h: F \otimes F \otimes F \rightarrow \bar{\mathbf{Q}}$ amounts to the choice of three non-trivial fifth roots of 1, $\omega_1, \omega_2, \omega_3$. For h to factor through $R \otimes \mathbf{Q}$ the right-hand identities in (1.5.2) force $\omega_i \omega_j \neq 1$, $i \neq j$. On the other hand, for the image of

$\pi^*\pi_*$ to be non-trivial, one must have $\omega_1\omega_2^2\omega_3^3 = 1$. One checks easily that the conditions

$$\omega_1\omega_2\omega_3 \neq 1,$$

$$\omega_i \neq 1, \quad \omega_i\omega_j \neq 1, \quad \omega_1\omega_2^2\omega_3^3 = 1, \quad i \neq j = 1,2,3$$

cannot all hold, so $\pi^*\pi_* = 0$. This proves the claim. □

Remarks The surface in (1.2) is called hyperelliptic; that in (1.5) is a Godeaux surface. The reader should be warned that the author has exercised his prerogative in selecting these examples very carefully. In particular, both examples have $P_g = 0$.

I want now to sketch Mumford's idea [4] for showing that $P_g > 0$ implies $T(X)$ is enormous. For another argument, see the appendix to this lecture. Choose a base point $p_0 \in X$ and define a map ρ_n from the nth symmetric product S^nX = set of unordered n-tuples of elements of X to $A_0(X)$, $\rho_n : S^nX \to A_0(X)$, where $\rho_n(x_1, \ldots, x_n) = \sum(x_i) - n(p_0)$. Notice if $T(X) = 0$, so $A_0(X) \cong \mathrm{Alb}(X)$, the fibres of ρ_n would be subschemes of S^nX of codimension $\leq q = \dim \mathrm{Alb}(X)$. Mumford proves

Theorem (1.6) *Assume $P_g > 0$ and let $t \in S^nX$ be a general point. Assume $t \in W \subset S^nX$ is a subscheme on which ρ_n is constant; that is, $\rho_n(W) = \rho_n(t)$. Then the codimension of W in S^nX is $\geq n$.*

Idea of proof Let ω be a non-zero global holomorphic 2-form on X. By pulling back along the various projections and adding up, one gets a 2-form

$$\tilde{\omega}_n = \mathrm{pr}_1^*\omega + \cdots + \mathrm{pr}_n^*\omega$$

on the cartesian product $X^n = X \times \cdots \times X$. The form $\tilde{\omega}_n$ is invariant under the action of the symmetric group S_n on X^n and so in a certain sense $\tilde{\omega}_n$ descends to a form ω_n on $S^nX = X^n/S_n$. (This part of the argument requires considerable care because S^nX will have singularities.) Quite precisely, if T is a nonsingular variety parameterizing an effective family of zero-cycles of degree n a general element of which consists of n distinct points with multiplicity 1, there will be a "pullback" $\omega_{n,T} \in \Gamma(T, \Omega_T^2)$ defined. Note that the singularities of S^nX occur at points $(x_1, \ldots, x_n)$ where two or more of the x_i coincide. On the complement $(S^nX)_{\mathrm{smooth}}$ of the singular set, ω_n is well defined. There will be an open set $T^0 \subset T$ and a morphism $T^0 \xrightarrow{\psi} (S^nX)_{\mathrm{smooth}}$, and $\omega_{n,T}$ will be a holomorphic extension of $\psi^*\omega_n$.

Using the definition of rational equivalence and the fact that there are no global holomorphic forms on projective space, Mumford shows that if the cycles in the family parameterized by T are all rationally equivalent, then $\omega_{n,T} = 0$. One next notices that if $t \in (S^n X)_{\text{smooth}}$ is general, the two-form ω_n will give a *non-degenerate* alternating pairing on the tangent space at t. Shifting t along W, our given subvariety, if necessary we may assume t is a nonsingular point of W. Resolving singularities of W then does not change the situation near t, but does show that our non-degenerate pairing, when restricted to the tangent space of W at t, gives zero. It follows that $\dim W \le \frac{1}{2}\dim S^n X = n$, so codim $W \ge n$. $\square$

Example (1.7) (Fatemi [10])   A surface with $P_g > 0$ and an interesting Chow group is the *Fano surface*. Let T be a cubic threefold, that is, a smooth hypersurface of degree 3 in $\mathbf{P}^4$. Let G be the Grassmann of lines in $\mathbf{P}^4$, and let $S \subset G$ be the subvariety of lines on T. The subvariety S is known to be a smooth connected surface. We write $\text{Pic}(S) = \text{CH}^1(S)$, and $\text{Pic}^0(S) = $ divisors algebraically equivalent to zero modulo rational equivalence. $\text{Pic}^0(S)$ is the group of closed points of the *Picard variety* of S, and there is an exact sequence

$$0 \to \text{Pic}^0(S) \to \text{Pic}(S) \to \text{NS}(S) \to 0,$$

where $\text{NS}(S) = $ Néron–Severi group of $S = $ subgroup of $H^2(S, \mathbf{Z})$ generated by classes of divisors.

Intersection of divisors gives bilinear maps

(1.7.1) $\text{Pic}(S) \otimes_{\mathbf{Z}} \text{Pic}^0(S) \to A_0(S),$

(1.7.2) $\text{Pic}^0(S) \otimes_{\mathbf{Z}} \text{Pic}^0(S) \to T(S).$

(If D^0 is algebraically equivalent to zero and D is any divisor, $D \cdot D^0$ will have degree 0. The fact that $D \cdot D^0 \in T(S)$ if both D and D^0 are algebraically equivalent to zero can be checked by noting that the map

$$\text{Pic}^0(S) \times \text{Pic}^0(S) \xrightarrow[\text{intersect}]{} A_0(S) \to \text{Alb}(S)$$

is a continuous bilinear map of compact tori, hence trivial (exercise!).)

Claim   *The intersection maps (1.7.1) and (1.7.2) are surjective.*

Proof   I will give the proof only for (1.7.1). The minor modification needed to prove (1.7.2) is left for the reader. Necessary facts about cubic 3-folds can be found in Clemens and Griffiths [11] or Tyurin [12]. Let $r, s \in S$ be general points. It will suffice (using divisibility of $A_0(S)$) to show that $2(r) - 2(s) \in$ Image($\text{Pic}^0(S) \otimes_{\mathbf{Z}} \text{Pic}(S) \to A_0(S)$). The line on T corresponding to $s \in S$ will be denoted ℓ_s, and D_s will denote the divisor on S

$$D_s = \text{Zariski closure of } \{t \in S \mid \ell_t \cap \ell_s \ne \varnothing, t \ne s\}.$$

In general, $s \notin D_s$, and $D_s \cdot D_t$ has degree 5. Thus, if r, s are general ("general," in this context, means the points in question do not lie on a finite number of *a priori* specified proper closed subvarieties of S) we can find a general t such that $\ell_s \cap \ell_t \neq \varnothing$ and $\ell_t \cap \ell_r \neq \varnothing$. Since $2((r) - (s)) = 2((r) - (t)) + 2((t) - (s))$, we may assume $\ell_r \cap \ell_s \neq \varnothing$.

Let L be the plane spanned by ℓ_r and ℓ_s in $\mathbf{P}^4$ and let ℓ_t be the third line in $L \cap T = \ell_r \cup \ell_s \cup \ell_t$.

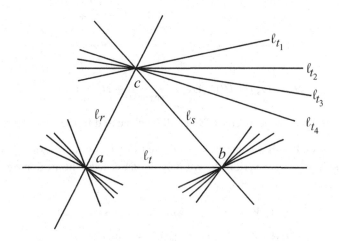

Through a general point of T there pass 6 lines, and we denote by $\theta : \mathrm{CH}_0(T) \to \mathrm{CH}_0(S)$ the correspondence thus defined. $\mathrm{CH}_0(T) \cong \mathbf{Z}$ (any two points on T can be connected by a chain of lines) so we have

$$\theta(a) = (r) + (t) + (s_1) + \cdots + (s_4)$$
$$\sim \theta(b) = (s) + (t) + (r_1) + \cdots + (r_4)$$
$$\sim \theta(c) = (r) + (s) + (t_1) + \cdots + (t_4).$$

On the other hand

$$D_r \cdot D_s = (t) + (t_1) + \cdots + (t_4),$$
$$D_s \cdot D_t = (r) + (r_1) + \cdots + (r_4),$$
$$D_r \cdot D_t = (s) + (s_1) + \cdots + (s_4).$$

Hence

$$\theta(a) - D_r \cdot D_t = (r) + (t) - (s),$$
$$\theta(b) - D_s \cdot D_t = (s) + (t) - (r),$$
$$(D_s - D_r) \cdot D_t \sim \theta(a) - D_r \cdot D_t - \theta(b) + D_s \cdot D_t = 2(r) - 2(s). \quad \square$$

Philosophy The dual to $\Gamma(S, \Omega_S^2)$ is the second cohomology group of the structure sheaf, $H^2(S, O_S)$ (Serre duality). The tangent space to $\mathrm{Pic}^0(S)$ is $H^1(S, O_S)$. In the case of the Fano surface, the cup product map induces an isomorphism $\Lambda^2 H^1(S, O_S) \cong H^2(S, O_S)$. In some mystical sense, the control exercised by the geometric genus P_g over the size of $T(S)$ can be seen in the following schema:

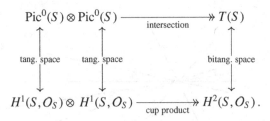

The idea of a bitangent space to $T(S)$ will be discussed further in Lecture 6.

I want to end with a conjecture, motivated in part by these examples. For any smooth projective variety V, we may define at least the beginnings of a filtration on $\mathrm{CH}_0(V)$ by

$$F^1 \mathrm{CH}_0(V) = A_0(V) = \mathrm{Ker}\,(\deg \colon \mathrm{CH}_0(V) \to \mathbf{Z}),$$

$$F^2 \mathrm{CH}_0(V) = \mathrm{Ker}\,(F^1 \mathrm{CH}_0(V) \to \mathrm{Alb}(V)).$$

Presumably the filtration will in general have n steps where $n = \dim V$. Except in certain simple cases like abelian varieties or products of curves (Bloch [13]), I don't have much feeling for what should come beyond F^2. We will therefore "truncate" at F^2.

Let X be a surface, and let Γ be a cycle on $V \times X$ with $\dim \Gamma = n = \dim V$. The cycle Γ then induces a map

$$\Gamma_* \colon \mathrm{CH}_0(V) \to \mathrm{CH}_0(X).$$

The above filtration being functorial for correspondences, we get also

$$\mathrm{gr}\,\Gamma_* \colon \mathrm{gr}\,\mathrm{CH}_0(V) \to \mathrm{gr}\,\mathrm{CH}_0(X) \cong \mathbf{Z} \oplus \mathrm{Alb}(X) \oplus T(X).$$

Conjecture (1.8) *The map* $\mathrm{gr}\,\Gamma_*$ *depends only upon the cohomology class* $[\Gamma] \in H^4(V \times X)$.

I want to discuss some consequences of this conjecture, and to try to make it more precise. Let $\Gamma(i, j) \in H^i(V) \otimes H^j(X)$ denote the Künneth components of Γ (since the graded pieces of CH_0 are either divisible or torsion free, we may as well work with rational cohomology). Because X is a surface, one knows (cf. Kleiman [15], 2A10 and 2.9) that the $\Gamma(i, j)$ are algebraic. Altering Γ by constant correspondences, we may suppose $\Gamma(0, 4) = \Gamma(4, 0) = 0$.

Lemma (1.9) *Assuming conjecture (1.8), $\Gamma(3, 1)_* : \mathrm{gr}\, \mathrm{CH}_0(V) \to \mathrm{gr}\, \mathrm{CH}_0(X)$ is zero. The map $\Gamma(1, 3)_*$ is zero on $F^2\, \mathrm{CH}_0(V)$.*

Proof Let $q = \dim \mathrm{Alb}(X)$. A Poincaré divisor D of Künneth type $(1,1)$ on $X \times \mathrm{Pic}^0(X)$ induces via correspondences an isomorphism

$$D_* : H^{2q-1}(\mathrm{Pic}^0(X)) \xrightarrow[\cong]{} H^1(X).$$

The inverse of this isomorphism will also be induced by an algebraic correspondence (Kleiman [15], 3.11), so we can factor $\Gamma(3, 1)_*$ via an algebraic correspondence $\Sigma \in H^3(V) \otimes H^{2q-1}(\mathrm{Pic}^0(X))$. Such a Σ is represented by an algebraic cycle on $V \times \mathrm{Pic}^0(X)$ of dimension $\dim V - 1$, so on the cycle level $\Sigma_* \mathrm{CH}_0(V) = (0)$. To study $\Gamma(1, 3)_*$ let $j : C \to X$ be a smooth hyperplane section. The map $\mathrm{Jacobian}(C) \to \mathrm{Alb}(X)$ is split up to isogeny (Poincaré reducibility). Since $H^3(X) \cong H^{2q-1}(\mathrm{Alb}(X))$ one shows that there exists an algebraic correspondence $\psi \in H^1(V) \otimes H^1(C)$ such that $(1 \times j_*)\psi = \Gamma(1, 3)$. Since $F^2\, \mathrm{CH}_0(C) = (0)$, this implies $\Gamma(1, 3)_* F^2\, \mathrm{CH}_0(V) = (0)$. $\square$

From the point of view of cycles, the interesting correspondences to study therefore are those of Künneth type $(2, 2)$. Looking at an example like (1.7) one might be tempted to formulate the fantastically vague:

Metaconjecture (1.10) *There is an equivalence of categories between a suitable category of polarized Hodge structures of weight 2 and a category built up from the $F^2\, \mathrm{CH}_0(X)$.*

Such an equivalence might provide a geometric interpretation of the category of Hodge structures of weight 2 analogous to the abelian variety interpretation for weight 1!

I want finally to record two other consequences of (1.8).

Proposition (1.11) *Assume conjecture (1.8) holds, and let X be a surface of geometric genus 0. Then $F^2\, \mathrm{CH}_0(X) = (0)$.*

Proof Let $\Delta \subset X \times X$ be the diagonal. Then $\Delta_* = \Delta(2,2)_* = \mathrm{Id}$ on $F^2\,\mathrm{CH}_0(X)$. Since X has geometric genus 0, $H^2(X)$ is generated by divisors, so $\Delta(2,2) = \sum a_{ij}\,D_i \times D_j$ where the D_i are divisors. But a zero-cycle can be moved away from a finite union of divisors, so $\Delta(2,2)_* : \mathrm{CH}_0(X) \to \mathrm{CH}_0(X)$ is zero.   □

Proposition (1.12) (Conjectural theorem of the hypersquare) *Assume conjecture (1.8) holds. Let X be a surface and $V = V_1 \times V_2 \times V_3$, with all varieties smooth and connected. Let Γ be a codimension-2 cycle on $V \times X$ and let $v_{ij} \in V_i$ be points, $i = 1,2,3$ and $j = 1,2$. Then*

$$\Gamma_*\Big(\sum_{j,k,\ell=1,2} (-1)^{j+k+\ell}(v_{ij}, v_{2k}, v_{3\ell}) \Big) = 0 \quad \text{in } \mathrm{CH}_0(X).$$

Proof View Γ as a family $\{\Gamma_t\}_{t \in V_3}$ of cycles on $V_1 \times V_2 \times X$ parameterized by V_3. The Γ_t are clearly all homologous, so

$$\Gamma_{v_{31}*} = \Gamma_{v_{32}*} : F^2\,\mathrm{CH}_0(V_1 \times V_2) \to F^2\,\mathrm{CH}_0(X).$$

But the cycle $\sum_{j,k=1,2}(-1)^{j+k}(v_{ij}, v_{2k})$ is an element of $F^2\,\mathrm{CH}_0(V_1 \times V_2)$ (exercise!). The desired identity is now straightforward.   □

For further discussion of this sort of conjecture, the reader can see Bloch [14].

References for Lecture 1

For general facts about algebraic surfaces over **C**, a good reference is

[1] A. Beauville, Surfaces algébriques complexes, *Astérisque*, **59** (1978),

as well as the references cited there.

Example (1.2) is done more generally in

[2] S. Bloch, A. Kas, and D. Lieberman, Zero cycles on surfaces with $P_g = 0$, *Compositio Math.*, **33** (1976), 135–145,

along with other surfaces with $P_g = 0$ and Kodaira dimension < 2.

Example (1.5) is due to H. Inose and M. Mizukami. Details and other analogous examples are in

[3] H. Inose and M. Mizukami, Rational equivalence of 0-cycles on some surfaces of general type with $p_g = 0$, *Math. Ann.*, **244** (1979), no. 3, 205–217.

To understand the role of the geometric genus, the reader should see the seminal paper of Mumford:

[4] D. Mumford, Rational equivalence of zero-cycles on surfaces, *J. Math. Kyoto Univ.*, **9** (1968), 195–204,

as well as

[5] A. A. Roitman, Γ-equivalence of zero-dimensional cycles (in Russian), *Mat. Sb. (N.S.)*, **86 (128)** (1971), 557–570. [Translation: Math USSR-Sb., **15** (1971), 555–567.]

[6] A. A. Roitman, Rational equivalence of zero-dimensional cycles (in Russian), *Mat. Sb. (N.S.)*, **89 (131)** (1972), 569–585, 671. [Translation: Math. USSR-Sb., **18** (1974), 571–588.]

[7] A. Mattuck, Ruled surfaces and the Albanese mapping, *Bull. Amer. Math. Soc.*, **75** (1969), 776–779.

[8] A. Mattuck, On the symmetric product of a rational surface, *Proc. Amer. Math. Soc.*, **21** (1969), 683–688.

For a more positive point of view, see

[9] S. Bloch, K_2 of Artinian **Q**-algebras with application to algebraic cycles, *Comm. Algebra*, **3** (1975), 405–428.

For a time it was believed that Mumford's technique could be applied to pluricanonical differentials, and that one could in this way show $A_0(X)$ was not representable for any non-ruled surface. A number of such statements appeared in the literature. The argument fails because the trace of a pluricanonical form under a finite ramified map can acquire singularities.

Example (1.7) was worked out by Fatemi.

[10] T. Fatemi, L'equivalence rationelle des zéro cycles sur les surfaces algébriques complexes a cup product surjectif, These du 3e cycle, Université de Paris VII (1979).

For more details about the geometry of the Fano surface and the cubic threefold, see

[11] C. H. Clemens and P. A. Griffiths, The intermediate Jacobian of the cubic threefold, *Ann. of Math. (2)*, **95** (1972), 281–356.

[12] A. N. Tyurin, Five lectures on three-dimensional varieties (in Russian), *Uspehi Mat. Nauk*, **27** (1972), no. 5, (167) 3–50. [Translation: Russian Math. Surveys, **27** (1972), no. 5, 1–53.]

Some references for the conjectures at the end of Lecture 1 are:

[13] S. Bloch, Some elementary theorems about algebraic cycles on abelian varieties, *Invent. Math.*, **37** (1976), 215–228.

[14] S. Bloch, An example in the theory of algebraic cycles, pp. 1–29 in *Algebraic K-Theory*, Lecture Notes in Math., no. 551, Springer, Berlin (1976).

[15] S. Kleiman, Algebraic cycles and the Weil conjectures, pp. 359–386 in *Dix exposés sur la cohomologie des schémas*, North Holland, Amsterdam (1968).

Appendix: On an argument of Mumford in the theory of algebraic cycles

S. Bloch[1]

Let X be a smooth projective surface over an algebraically closed field k. Let $CH_0(X)$ denote the Chow group of zero cycles modulo rational equivalence on X, and let $A_0(X) \subset CH_0(X)$ be the subgroup of cycles of degree 0. I will say that $A_0(X)$ is finite dimensional if there exists a complete smooth (but possibly disconnected) curve C mapping to X such that the map $J(C) = \text{Jacobian}(C) \to A_0(X)$ is surjective. Some years ago, Mumford proved, in the case $k = \mathbf{C}$, that $P_g(X) > 0$ implies $A_0(X)$ is not finite dimensional. The purpose of the present note is to prove an analog of this result applicable in all characteristics. The role of the geometric genus, which is not a good invariant in characteristic p, is played by the "transcendental part" of $H^2_{\text{et}}(X, \mathbf{Q}_\ell)$. The present proof also reveals the influence of the finite dimensionality of the Chow hypothesis on the structure of the "motive" of X.

The idea that one could deduce interesting information about the Chow group by considering the generic zero-cycle was suggested by Colliot-Thélène. I am indebted to him for letting me steal it.

Lemma (1A.1) *Let X be a smooth variety over an algebraically closed field k, and Y any k-variety. Let $n \geq 0$ be an integer. Then, writing $K = k(Y)$,*

$$CH^n(X_K) \cong \varinjlim_{U \subset Y \text{ open}} CH^n(X \times_k U),$$

where CH^n equals codimension-n cycles modulo rational equivalence.

Proof For any variety W and any integer m, let W^m = set of points of codi-

---

[1] The author gratefully acknowledges support from the NSF and from the conference on Chow groups of rational varieties. The present note grew out of conversations with Colliot-Thélène at the conference.

mension m on W. One has

$$(X_K)^m = \varprojlim_U (X \times_k U)^m.$$

Since

$$\mathrm{CH}^n(X_K) = \mathrm{Coker}\Big(\coprod_{x \in (X_K)^{n-1}} K(x)^* \longrightarrow \coprod_{x \in (X_k)^n} \mathbf{Z} \Big),$$

$$\mathrm{CH}^n(X \times U) = \mathrm{Coker}\Big(\coprod_{y \in (X \times U)^{n-1}} k(x)^* \longrightarrow \coprod_{y \in (X \times U)^n} \mathbf{Z} \Big),$$

the desired result is immediate.   □

Proposition (1A.2)   *Let X be a smooth projective surface over k and let $\Omega \supset k$ be a universal domain in the sense of Weil. Assume $A_0(X_\Omega)$ is finite dimensional. Then there exist one-dimensional subschemes $C', C'' \subset X$ and a 2-cycle Γ supported on $(C' \times X) \cup (X \times C'')$ such that some non-zero multiple of the diagonal Δ on $X \times_k X$ is rationally equivalent to Γ.*

Proof   Let $C \to X$ be such that $J(C_\Omega) \twoheadrightarrow A_0(X_\Omega)$, and let $C' \subset X$ be the image of C. Enlarging k, we may assume C' defined over k.

Lemma (1A.3)   *Let $k \subset K \subset K'$ be extensions of fields. Then the kernel of $\mathrm{CH}^2(X_K) \to \mathrm{CH}^2(X_{K'})$ is torsion.*

Proof   If $[K' : K] < \infty$ this follows from the existence of a norm $\mathrm{CH}^2(X_{K'}) \to \mathrm{CH}^2(X_K)$. The case K' algebraic over K follows by a limit argument. Enlarging K and K', we may thus assume K algebraically closed. In this case, $\mathrm{CH}^2(X_{K'})$ is a limit of Chow groups $\mathrm{CH}^2(X \times_K U)$, where U is a K-variety of finite type. A K-point of U gives a section of $\mathrm{CH}^2(X) \to \mathrm{CH}^2(X \times_K U)$ so the lemma follows.   □

Proof of proposition (1A.2)   Let K be the function field of X over k, and fix an embedding $K \hookrightarrow \Omega$. Let $P \in X(\Omega)$ be the corresponding point. Our hypotheses imply $\mathrm{CH}^2((X - C')_\Omega) = (0)$, so by Lemma (1A.3), there exists $N \geq 1$ such that $N(P) = 0$ in $\mathrm{CH}^2((X - C')_K)$. (We abuse notation by writing P also for the generic point of X. In other words, P is the image in $\mathrm{CH}^2(X_K)$ of the diagonal Δ in $X \times X$.) By Lemma (1A.1), there exists $U \subset X$ open $\neq \varnothing$ such that $N \cdot \Delta$ is rationally equivalent to zero on $(X - C') \times_K U$. Let $C'' \subset X$ be a subscheme of codimension 1 containing $X - U$. The proposition now follows from the exact sequence

$$\left.\begin{cases} \text{cycles supported on} \\ (C' \times X) \cup (X \times C'') \end{cases}\right\} \to \mathrm{CH}^2(X \times_k X)$$

$$\to \mathrm{CH}^2((X - C') \times_k (X - C'')) \to 0. \qquad \square$$

Exercise (1A.4) Generalize the definition of finite dimensionality for $A_0(X)$ to varieties X of dimension > 2 and prove the analogue of Proposition (1A.2).

Fix now an ℓ prime to char p, and define

$$H^2(X)_{\text{trans}} = H^2_{\text{et}}(X, \mathbf{Q}_\ell)/\text{Image}(\text{NS}(X) \otimes \mathbf{Q}_\ell \longrightarrow H^2_{\text{et}}(X, \mathbf{Q}_\ell)),$$

where $\text{NS}(X)$ is the Néron–Severi group. We systematically ignore twisting by roots of 1. An alternative description of $H^2(X)_{\text{trans}}$ is given by

$$H^2(X)_{\text{trans}} = \text{Image}(H^2_{\text{et}}(X, \mathbf{Q}_\ell) \to H^2_{\text{Gal}}(\bar{K}/K, \mathbf{Q}_\ell)).$$

Note $H^2_{\text{et}}(X)$, $\text{NS}(X)$, and hence also $H^2(X)_{\text{trans}}$ are functorial for correspondences, which we will take to be elements in $\mathrm{CH}^2(X \times_k X)$. In particular, the diagonal induces the identity on $H^2(X)_{\text{trans}}$.

Lemma (1A.5) *With notation as in Proposition (1A.2), let Γ be a codimension-2 cycle on $X \times X$ supported on $(C' \times X) \cup (X \times C'')$. Then the correspondence $\Gamma_* : H^2(X)_{\text{trans}} \to H^2(X)_{\text{trans}}$ is zero.*

Proof It is convenient to work with étale homology, which is defined for any k-scheme Y which can be embedded in a smooth k-variety. Write $\Gamma = \Gamma' + \Gamma''$ with $\text{Supp}\,\Gamma' \subset C' \times X$ and $\text{Supp}\,\Gamma'' \subset X \times C''$. Let $i' : C' \times X \hookrightarrow X \times X$, $i'' : X \times C'' \hookrightarrow X \times X$, and write $[\Gamma'] \in H_4(C' \times X, \mathbf{Q}_\ell)$, $[\Gamma''] \in H_4(X \times C'', \mathbf{Q}_\ell)$. For $\alpha \in H^2_{\text{et}}(X, \mathbf{Q}_\ell)$, we have

$$\Gamma'_* \alpha = \text{pr}_{2*}(\text{pr}_1^*(\alpha) \cdot i'_*[\Gamma']) = \text{pr}_{2*} i'_*(i'^* \text{pr}_1^*(\alpha) \cdot \Gamma')$$

$$= \text{pr}_{2*} i'_*(\text{pr}_1^* i'^*(\alpha) \cdot \Gamma),$$

with morphisms labeled as indicated:

$$\begin{array}{ccc} C' \times X & \stackrel{i'}{\lhook\joinrel\longrightarrow} & X \times X \\ {\scriptstyle \text{pr}_1} \downarrow & & \downarrow {\scriptstyle \text{pr}_1} \\ C' & \stackrel{i'}{\lhook\joinrel\longrightarrow} & X. \end{array}$$

Note that $\text{NS}(X) \otimes \mathbf{Q}_\ell$ is self-dual under the cup product pairing on $H^2_{\text{et}}(X, \mathbf{Q}_\ell)$. Without changing the image of α in $H^2(X)_{\text{trans}}$ we may assume, therefore, that α is perpendicular to $\text{NS}(X)$. Since $H_2(C', \mathbf{Q}_\ell)$ is generated by the classes of components of C', we find $i'^*(\alpha) = 0$ in $H^2(C', \mathbf{Q}_\ell)$, so $\Gamma'_*(\alpha) = 0$.

It remains to show $\Gamma''_*(\alpha) = 0$ where Γ'' is supported on $X \times C''$. We have a diagram

$$
\begin{array}{ccc}
X \times C'' & \xrightarrow{\ i''\ } & X \times X \\
\Big\downarrow{\scriptstyle \mathrm{pr}_2} & & \Big\downarrow{\scriptstyle \mathrm{pr}_2} \\
C'' & \xrightarrow{\ i''\ } & X
\end{array}
$$

and the projection formula gives

$$
\begin{aligned}
\Gamma''_*\alpha &= \mathrm{pr}_{2*}(i''_*\Gamma'' \cdot \mathrm{pr}_1^*\alpha) \\
&= i''_*\mathrm{pr}_{2*}(\Gamma'' \cdot i''^*\mathrm{pr}_1^*\alpha) \\
&\in \mathrm{Image}(H_2(C'', \mathbf{Q}_\ell) \to H^2(X, \mathbf{Q}_\ell)) \subset \mathrm{NS}(X) \otimes \mathbf{Q}_\ell,
\end{aligned}
$$

so $\Gamma''_*\alpha \to 0$ in $H^2(X)_{\mathrm{trans}}$. $\qquad\qquad\qquad\qquad\qquad\qquad\qquad\qquad\square$

We have proven:

Theorem (1A.6) *Let X be a smooth projective surface over an algebraically closed field k, and let $k \subset \Omega$ be a universal domain. If $H^2_{\mathrm{et}}(X, \mathbf{Q}_\ell) \neq \mathrm{NS}(X) \otimes \mathbf{Q}_\ell$ ($\ell \neq \mathrm{char}\, k$), then $A_0(X_\Omega)$ is not finite dimensional.*

Exercise (1A.7) Formulate and prove an analogous result for varieties of dimension > 2.

Question (1A.8) If E is a supersingular elliptic curve over a field k of characteristic $\neq 0, \ell$, then $H^2_{\mathrm{et}}(E \times E, \mathbf{Q}_\ell) = \mathrm{NS}(E \times E) \otimes \mathbf{Q}_\ell$. Is $A_0(E \times E)$ finite dimensional?

When $k = \mathbf{C}$, $H^2_{\mathrm{et}}(X) \neq \mathrm{NS}(X) \otimes \mathbf{Q}_\ell$ if and only if $P_g(X) > 0$, so we recover Mumford's result:

Corollary (1A.9) *When $k = \mathbf{C}$, $P_g(X) > 0$ implies $A_0(X_{\mathbf{C}})$ is not finite dimensional.*

References for Lecture 1 Appendix

D. Mumford, Rational equivalence of 0-cycles on surfaces, *J. Math. Kyoto Univ.*, **9** (1968), 195–204.

A. A. Roitman, Rational equivalence of zero-dimensional cycles (in Russian), *Mat. Sb. (N.S.)*, **89 (131)** (1972), 569–585, 671. [Translation: Math. USSR-Sb., **18** (1974), 571–588.]

2

Curves on threefolds and intermediate jacobians

The purpose of this lecture is to give some feeling for the geometry of curves on threefolds over the complex numbers; in particular the link with *intermediate jacobians*. We will focus on the example of a quartic threefold, a smooth hypersurface X in $\mathbf{P}^4$ of degree 4 and will show in this case that $A^2(X)$ is isomorphic to the intermediate jacobian $J^2(X)$. (In all honesty the proof will not be quite general, as we shall suppose for technical simplicity that the curve of lines on X is smooth. This is false, for example, for the Fermat quartic; see Tennison [9]. No one has worked out the details in the general case, which would involve jacobians of singular curves. It seems likely the final result would be the same.)

The intermediate jacobians to which these cycles are related can be described as follows. Let X be a smooth projective variety and $r > 0$ an integer. The complex cohomology $H^{2r-1}(X, \mathbf{C})$ has a Hodge filtration (Deligne [5])

$$H^{2r-1}(X, \mathbf{C}) = F^0 H^{2r-1} \supset F^1 \supset \cdots \supset F^{2r-1} \supset (0).$$

This complex vector space also has a $\mathbf{Z}$-structure defined by the image of $H^{2r-1}(X, \mathbf{Z}) \to H^{2r-1}(X, \mathbf{C})$. In particular, it has an $\mathbf{R}$-structure, so it makes sense to talk about conjugating an element or a subspace. The key property of the Hodge filtration is

$$F^i \oplus \overline{F^{2r-i}} \cong H^{2r-1}(X, \mathbf{C}),$$

$$F^i \cap \overline{F^{2r-i-1}} \cong H^{i,2r-i-1},$$

where

$$H^{i,j} \cong H^j(X, \Omega_X^i).$$

In particular, we see from this that

$$\dim_{\mathbf{C}} F^r = \tfrac{1}{2} \dim_{\mathbf{C}} H^{2r-1}(X, \mathbf{C}),$$

$$F^r \cap \operatorname{Image}(H^{2r-1}(X, \mathbf{Z}) \to H^{2r-1}(X, \mathbf{C})) = (0),$$

so the quotient

$$J^r(X) = H^{2r-1}(X, \mathbf{C})/F^r + H^{2r-1}(X, \mathbf{Z})$$

is a compact complex torus, called the intermediate jacobian.

If $\dim X = n$, Poincaré duality implies

$$H^{2r-1}(X, \mathbf{C})/F^r \cong (F^{n-r+1} H^{2n-2r+1}(X, \mathbf{C}))^*,$$

where * denotes $\mathbf{C}$-linear dual. It follows that

$$J^r(X) \cong (F^{n-r+1} H^{2n-2r+1}(X, \mathbf{C}))^* / H_{2n-2r+1}(X, \mathbf{Z}).$$

For example, if $r = n$ we find

$$J^n(X) \cong \Gamma(X, \Omega_X^1)^* / H_1(X, \mathbf{Z}) \cong \operatorname{Alb}(X).$$

The next step is to define cycle classes in these intermediate jacobians. In this lecture we restrict ourselves to a concrete intuitive description valid for the case at hand.

Let X be a smooth projective threefold, and let $\gamma = \sum n_i \gamma_i$ be an algebraic 1-cycle on X. We assume γ is *homologous* to 0, that is $[\gamma] = 0$ in $H^4(X, \mathbf{Z})$, and the idea will be to mimic the construction in Lecture 1 of $X \to \operatorname{Alb}(X)$ by associating to $(p) - (p_0)$ the integral

$$\int_{p_0}^{p} \in \Gamma(X, \Omega_X^1)^* / H_1(X, \mathbf{Z}).$$

We choose a 3-chain Δ on X with $\partial \Delta = \gamma$ and consider $\int_{\Delta}$ as an element in the dual space to the space of 3-forms of types $(3, 0) + (2, 1)$. (We are thinking of $J^2(X) = F^2 H^3(X, \mathbf{C})^* / H_3(X, \mathbf{Z})$.) The first point to note is that if ω is of type $(3, 0) + (2, 1)$ and ω is exact, then we can write $\omega = d\eta$, where η has type $(2, 0)$. (This is an easy consequence of Hodge theory; see Griffiths and Harris [7].) Stokes' theorem gives

$$\int_{\Delta} \omega = \int_{\gamma} \eta = 0$$

because there are no $(2, 0)$ forms on a curve. Notice finally if Δ' is another 3-chain with $\partial \Delta' = \partial \Delta = \gamma$, then $\Delta' - \Delta$ is a 3-cycle so

$$\int_{\Delta'} - \int_{\Delta} \in \operatorname{Image}(H_3(X, \mathbf{Z}) \to F^2 H^3(X, \mathbf{C})^*).$$

These remarks suffice to define a cycle class

$$j(\gamma) \in J^2(X).$$

Recall that $A^2(X)$ is the group of codimension-2 cycles algebraically equivalent to zero, modulo rational equivalence. Since a cycle algebraically equivalent to zero is certainly homologous to zero, one can hope for a map

$$\Theta: A^2(X) \to J^2(X), \qquad \Theta(\gamma) = j(\tilde{\gamma}),$$

where $\tilde{\gamma}$ is a codimension-2 cycle representing $\gamma \in A^2(X)$.

Lemma (2.1) *The above map Θ is well defined. That is, given a family $\{\Gamma_t\}_{t \in \mathbf{P}^1}$ of codimension-2 cycles parameterized by $\mathbf{P}^1$, $j(\Gamma_t) \in J^2(X)$ is independent of t.*

Proof It is possible to show that $t \mapsto j(\Gamma_t)$ defines a holomorphic function $\mathbf{P}^1 \to J^2(X)$. Since $\mathbf{P}^1$ is simply connected, this lifts to a holomorphic map $\mathbf{P}^1 \to \mathbf{C}^N$, where $N = \dim J^2(X)$. But $\mathbf{P}^1$ has no non-constant global holomorphic functions, so any such map is constant. □

We want to study the map Θ when $X \subset \mathbf{P}^4$ is a smooth hypersurface of degree 4. The family of lines on X is known to form a connected curve F (Bloch and Murre [3]). In order to avoid certain technical problems which have never been carefully considered, we will assume that F is smooth. This is the case "in general" but not, for example, for $X : T_0^4 + T_1^4 + T_2^4 + T_3^4 + T_4^4 = 0$ (Tennison [9]).

Theorem (2.2) *Let X be a smooth quartic threefold and assume the curve of lines F is smooth. Then*

$$\Theta: A^2(X) \to J^2(X)$$

is an isomorphism.

Remark (2.3) If $s \in F$ corresponds to the line $\ell_s \subset X$, one can define an *incidence correspondence* Σ on F essentially by taking the closure of the set

$$\{(s, t) \in F \times F - \Delta \mid \ell_s \cap \ell_t \neq \varnothing\}.$$

One can then identify $J^2(X)$, using a beautiful idea of Tyurin [10] (note, however, the criticism of Tyurin's argument in Bloch and Murre [3]) with *generalized Prym*

$$P_\Sigma(F) = \text{Image}\,(\Sigma_* - 1 : J(F) \to J(F)).$$

Here $J(F)$ is the jacobian of F and Σ_* satisfies a certain quadratic relation $\Sigma_*^2 + (q - 2)\Sigma_* - (q - 1) = 0$. The standard Prym would correspond to the case

$q = 1$. For the quartic threefold, $q = 24$. We shall not pursue these ideas further here.

As a first step toward proving (2.2), we show

Proposition (2.4) *Let X be a smooth quartic threefold. Then* $\Theta \colon A^2(X) \to J^2(X)$ *is an isogeny (i.e.,* Θ *is surjective with finite kernel).*

Proof In several lemmas:

Lemma (2.5) *Let X be a smooth projective threefold, and let* $\pi \colon Y \to X$ *be obtained by blowing up a non-singular closed subvariety* $V \subset X$. *Then* $\Theta_X \colon A^2(X) \to J^2(X)$ *is an isogeny if and only if* Θ_Y *is.*

Proof If V is a point, then $A^2(X) \cong A^2(Y)$ and $J^2(X) \cong J^2(Y)$. If V is a curve, one finds (cf. Beauville [1], Clemens and Griffiths [4]) that $A^2(Y) \cong A^2(X) \times J(V)$ and $J^2(Y) \cong J^2(X) \times J(V)$, where $J(V)$ is the jacobian. In either case, the assertion is clear since Θ_Y induces the identity on $J(V)$. □

Lemma (2.6) *Let X and Y be smooth projective threefolds, and let* $\pi \colon Y - \to X$ *be a rational map of degree* $d \neq 0$ *(i.e.* $[k(Y) : k(X)] = d$). *If* Θ_Y *is an isogeny, then so is* Θ_X.

Proof One knows from resolution theory that there exists a map $Y' \to Y$ obtained by a succession of blowings up with nonsingular centers such that in the diagram

$$
\begin{array}{c}
Y' \\
\Big\downarrow \searrow {\scriptstyle \pi'} \\
Y \underset{\pi}{-} \to X
\end{array}
$$

the arrow π' is everywhere defined. We know that Θ_Y an isogeny implies $\Theta_{Y'}$ an isogeny, so we are reduced to the case $\pi \colon Y \to X$ everywhere defined.

Consider the diagram (commutative for either π_* or π^* – see Lieberman [12])

$$
\begin{array}{ccc}
A^2(Y) & \xrightarrow{\ \Theta_Y\ } & J^2(Y) \\
{\scriptstyle \pi_*}\Big\updownarrow{\scriptstyle \pi^*} & & {\scriptstyle \pi_*}\Big\updownarrow{\scriptstyle \pi^*} \\
A^2(X) & \xrightarrow{\ \Theta_X\ } & J^2(X).
\end{array}
$$

We have $\pi_*\pi^* = $ multiplication by d, and since $A^2(X)$ and $J^2(X)$ are divisible (Lecture 1, (1.3)),

$$
\pi_* \colon \pi^*(A^2(X)) \twoheadrightarrow A^2(X), \qquad \pi_* \colon \pi^*(J^2(X)) \twoheadrightarrow J^2(X).
$$

In particular, Θ_X is surjective. This implies that $\pi^*(A^2(Y)) \twoheadrightarrow \pi^*(J^2(X))$, and hence that this map is an isogeny (its kernel is contained in $\operatorname{Ker}\Theta_Y$). The map $\pi_*: \pi^*(J^2(X)) \twoheadrightarrow J^2(X)$ is a map of tori and hence easily seen to be an isogeny, so the composition $\Theta_X \circ \pi_* = \pi_* \circ \Theta_Y: \pi^*(A^2(X)) \to J^2(X)$ is an isogeny. It now follows that Θ_X is an isogeny. $\qquad\square$

A rational map $\pi: V -- \to W$ will be called a *conic bundle* if there exists an open dense set $W^0 \subset W$ for the Zariski topology such that π is everywhere defined on $\pi^{-1}(W^0)$, and a closed embedding

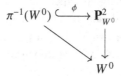

realizing $\pi^{-1}(W^0)$ as a family of conic curves over W^0. If $\bar\eta \to W$ denotes the geometric generic point in the sense of algebraic geometry, this is equivalent to the condition that the fibre $V_{\bar\eta}$ be isomorphic to the projective line $\mathbf{P}^1_{\bar\eta}$.

Lemma (2.7) *Let $\pi: V \to W$ be a conic bundle, where V and W are smooth and projective of dimensions three and two respectively. Assume that the map $A_0(W) \to \operatorname{Alb}(W)$ is an isogeny. Then $A^2(V) \xrightarrow{\Theta_V} J^2(V)$ is an isogeny.*

Proof Since the geometric generic fibre of π is $\mathbf{P}^1$, there exists a map $f: W' \to W$ of finite degree and a diagram

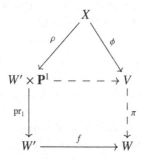

where ρ is obtained by a succession of blowings up with nonsingular centers, and ϕ is everywhere defined.

We now enumerate a series of exercises for the reader.

Exercise 1 Let $T(W') = \operatorname{Ker}(A_0(W') \to \operatorname{Alb}(W'))$. Show that $f_*T(W') = 0$.

Exercise 2

$$T(W') \xrightarrow[\cong]{f^*} \operatorname{Ker}(A^2(W' \times \mathbf{P}^1) \to J^2(W' \times \mathbf{P}^1)) \xrightarrow[\cong]{\rho^*} \operatorname{Ker}(A^2(X) \to J^2(X)).$$

Exercise 3 $\pi^* \circ f_* = \phi_* \circ \rho^* \circ \mathrm{pr}_1^* : A_0(W') \to A^2(V)$. (Hint: Move to general position.)

Exercise 4 $\phi_* : \mathrm{Ker}\,(A^2(X) \to J^2(X)) \twoheadrightarrow \mathrm{Ker}\,(A^2(V) \to J^2(V))$.

Exercise 5 Prove (2.7). □

We return now to the proof of (2.4). There exists a non-empty open subset X^0 of our quartic threefold X such that for all $x \in X^0$ we have:

(i) The intersection of the tangent hyperplane H_x to X at x with X, $H_x \cap X$, has an isolated ordinary double point at x and no other singularities.

(ii) There exist no lines $\ell \subset \mathbf{P}^4$ supported on V and passing through x. (This is because the family of lines on X has dimension 1.)

For $x \in X^0$, let $Q_x \subset H_x$ be the tangent cone to x on $H_x \cap X$. Set-theoretically Q_x is the union of all lines $\ell \subset \mathbf{P}^4$ which are at least triply tangent to X at x. In terms of equations, one can choose homogeneous forms $T_0, \ldots, T_4$ on $\mathbf{P}^4$ such that $x = (1, 0, 0, 0, 0)$ and $H_x : T_4 = 0$. The fact that x is an ordinary double point on $X \cap H_x$ means that the equation for X has the form

$$T_0^2 \cdot q_2\,(T_1, T_2, T_3) + T_0 \cdot r_3(T_1, \ldots, T_4)$$
$$+\, s_4\,(T_1, \ldots, T_4) + T_4 \cdot t_3\,(T_0, \ldots, T_4)\,,$$

where q_2, r_3, s_4, t_3 denote homogeneous polynomials of degrees 2, 3, 4, and 3 respectively. Moreover, q_2 is non-degenerate (i.e. equivalent after change of coordinates to the quadric $T_1^2 + T_2 T_3$), and $Q_x : T_4 = q_2\,(T_1, T_2, T_3) = 0$. In particular, Q_x is a cone over the smooth rational curve D_x defined by

$$D_x : T_0 = T_4 = q_2\,(T_1, T_2, T_3) = 0\,.$$

There is a conic bundle over X, $\pi : D \to X$ defined with $\pi^{-1}(x) = D_x$ for $v \in X^0$.

There is a rational morphism $\rho : D \to X$ defined as follows: for $x \in X^0$ and $d \in D_x$, let $\ell_d \subset Q_x$ be the line (ruling) passing through d and x. The line ℓ_d meets X in 4 points, at least three of which coincide with x. Define $\rho(d) =$ "fourth point of intersection of ℓ_d and X."

Lemma (2.8) *Let $x \in X^0$ and define $C_x = \rho(D_x)$. Then $C_x = Q_x \cap X$ is a reduced, irreducible rational curve of degree 8. If $x' \in X^0, x' \neq x$, then $C_{x'} \neq C_x$.*

Proof of lemma $Q_x \cap X = Q_x \cap (X \cap H_x)$ is the intersection of two distinct irreducible hypersurfaces in H_x and hence has pure dimension 1. In particular,

$\{x\} \in Q_x \cap X$ is not an isolated component. It is clear, set-theoretically, that $\rho(D_x) \subset Q_x \cap X$ and $Q_x \cap X - \{x\} \subset \rho(D_x)$, so we have $\rho(D_x) = Q_x \cap X$.

The degrees of Q_x and X are 2 and 4 respectively, so $\deg C_x = 8$. Also $C_x = \rho(D_x)$, so C_x is an irreducible rational curve. To show that the intersection $Q_x \cap X$ has multiplicity 1, choose two general points $p_1, p_2 \in C_x$ and let $\ell_1, \ell_2 \subset Q_x$ be the corresponding lines. There exists a hyperplane $L \subset \mathbf{P}^4$ such that $L \cap Q_x = \ell_1 \cup \ell_2$. Note that $(\ell_1 \cup \ell_2) \cap X$ contains p_1, p_2 with multiplicity one, because $\ell_i \cap X$ contains x with multiplicity 3. On the other hand,

$$(\ell_1 \cup \ell_2) \cap X = L \cap Q_x \cap X = L \cap C_x.$$

If the intersection $Q_x \cap X$ were not smooth at p_i, the multiplicity of p_i on $L \cap Q_x \cap X$ would be > 1.

It remains to show that $x' \in X^0$, $x' \neq x$, implies $C_{x'} \neq C_x$. Note that $C_x \subset H_x$, $C_{x'} \subset H_{x'}$, $H_x \neq H_{x'}$. If $C_x = C_{x'}$, we would have $C_x \subset Q_x \cap H_{x'}$ = curve of degree 2. This is impossible (even set-theoretically) by reason of degrees. □

Returning to the proof of (2.4), let $\pi^0 : D^0 \to X^0$ denote the restriction of π to $D^0 = \pi^{-1}(X^0)$. The morphism π^0 is a regular morphism, as is $\rho^0 : D^0 \to X$. The idea now is to construct a surface S and a conic bundle U over S by a "bootstrap" technique. Let $\Gamma \subset X$ be a reduced, irreducible curve. We assume $\Gamma^0 = \Gamma \cap X^0 \neq \emptyset$, and that for some $x \in \Gamma^0$, $\Gamma \neq C_x$. It follows from these hypotheses that $S^0 = \pi^{0-1}(\Gamma^0)$ is an irreducible quasi-projective surface, and $\rho^0(S^0) \subset X$ has dimension 2. Note that $\Gamma^0 \subset \rho^0(S^0)$ (because $x \in C_x = \rho(D_x)$), so $\rho^0(S^0) \cap X^0 \neq \emptyset$. Choose a projective desingularization and completion S of S^0 such that $\rho^0|_{S^0}$ extends to a morphism $\psi : S \to X$. Define U^0 by the fibre square

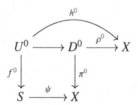

$U^0 \neq \emptyset$, and we have by composition a regular morphism $h^0 : U^0 \to X$.

S is generically fibred over Γ with rational fibres, so it is easy to see $A^2(S) \cong$ Alb(S). Also $f^{0-1}(s) \cong \mathbf{P}^1$ for $s \in S$ a general point. The hypotheses of (2.7) are thus verified (taking U = suitable completion of U^0) and (2.4) will follow if we show $h^0(U^0)$ is dense in X.

Suppose that $h^0(U^0)$ is not dense in X. This can only happen if for all points $x \in \psi(S) \cap X^0$, $C_x \subset \psi(S)$. Take another curve Γ' analogous to Γ such that $\Gamma' \not\subset \psi(S)$ but $\Gamma' \cap \psi(S) \cap X^0 \neq \emptyset$. Now repeat the above construction replacing

Γ by Γ'!! We get $\psi': S' \to X$. Every surface in X is a hypersurface section by Lefschetz theory, so $\psi'(S') \cap \psi(S) =$ finite union of curves. Let $E = X^0 \cap \psi'(S') \cap \psi(S)$. E is non-empty (it meets Γ') and open in $\psi(S) \cap \psi'(S')$, hence E is an infinite set.

Assume now that the triple (U', S', f') is no good either – that is, that $\dim h'(U'^0) < 3$. Then for every $x \in E$, we must have $C_x \subset \psi(S) \cap \psi'(S')$. This intersection is only a finite union of curves, so the infinite set E must contain points $x \neq x'$ with $C_x = C_{x'}$. This contradicts (2.8), so we conclude that one of the maps $h: U \to X$ or $h': U' \to X$ must be of finite degree. This completes the proof of (2.4).   □

We now have for the quartic threefold an isogeny $\theta_X: A^2(X) \approx J^2(X)$, and we must verify

Theorem (2.2) Θ_X *is an isomorphism.*

Proof Let $V \supset X$ be a quartic *fourfold* which is general containing X. Let $C \subset S$ be respectively the curve of lines on X and the threefold of lines on V. One knows that S and C are connected (for details cf. Bloch and Murre [3], §1) and we have assumed C is smooth. Writing $\mathrm{Gr}(1, n)$ for the Grassmann of lines in $\mathbf{P}^n$ we have $\mathrm{Gr}(1, n)$ is locally a complete intersection (in fact, smooth) of codimension 2 in $\mathrm{Gr}(1, n+1)$. Since scheme-theoretically $C = S \cap \mathrm{Gr}(1, 4) \subset \mathrm{Gr}(1, 5)$, smoothness of C implies smoothness of S along C. We now resolve singularities of S, pulling back the line correspondence in $S \times V$, and assume both C and S smooth:

(2.9)

where $L(C)$ and $L(S)$ are the incidence correspondences

$$L(?) = \{(a, b) \mid a \in ?, b \in \ell_a = \text{ line corresponding to } a\}$$

and

$$P = L(S) \times_V X = \{(a, b) \mid a \in S, b \in \ell_a \cap X\}.$$ □

Lemma (2.10)   *P is isomorphic to the blow-up of S along C.*

Proof One checks easily that P is smooth, p_1 is an isomorphism off the pre-image of C, and $p_1^{-1}(C)$ is a Cartier divisor. By a universal property of blowings-up, this gives a map

The map π is bijective and birational. Since $\mathrm{BL}_S(C)$ is smooth, π is an isomorphism. $\qquad\square$

From the known cohomological structure of blowings-up, we get

$$(2.11) \qquad H^3(P,\mathbf{Z}) \cong H^3(S,\mathbf{Z}) \oplus H^1(C,\mathbf{Z}).$$

Lemma (2.12) *The factor $H^3(S,\mathbf{Z}) \subset H^3(P,\mathbf{Z})$ maps to zero under the map $p_{2*}: H^3(P,\mathbf{Z}) \to H^3(X,\mathbf{Z})$.*

Proof $H^3(S)$ sits in $H^3(P)$ via $p_1^* = j^*q_1^*$. Since $p_{2*}j^* = k^*q_{2*}$, it suffices to show that $q_{2*}q_1^*H^3(S) = 0$. But this group sits in $H^3(V,\mathbf{Z}) = (0)$ by Lefschetz theory. $\qquad\square$

Lemma (2.13) *The map $p_{2*}: H^3(P,\mathbf{Z}) \to H^3(X,\mathbf{Z})$ is surjective.*

Before giving the proof of this, let me show how it implies the theorem. First, by (2.11) and (2.12) we see the correspondence

$$r_{2*}r_1^*: H^1(C,\mathbf{Z}) \to H^3(X,\mathbf{Z})$$

is surjective, so the corresponding map on jacobians

$$J(C) \to J^2(X)$$

has connected fibres. The diagram

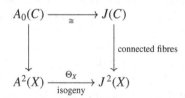

shows that Θ_X is an isogeny with connected fibres, hence an isomorphism, proving (2.2).

For the proof of (2.13), we work with homology rather than cohomology. Recall that given a Lefschetz pencil, $\{X_t\}_{t\in\mathbf{P}^1}$, of m-dimensional hyperplane sections of a smooth projective variety V and a base point $0 \in \mathbf{P}^1$ with $X = X_0$ smooth, there is associated to any choice of a singular fibre X_{t_i} and a path ℓ

from 0 to t_i on $\mathbf{P}^1$ a cycle class $\gamma \in H_m(X, \mathbf{Z})$. The class γ is the base (boundary) of a sort of cone Γ supported on $\bigcup_{t \in \ell} X_t$ with vertex the singular point on X_{t_i} (Wallace [11]). These vanishing cycles are known to generate the kernel of $H_m(X, \mathbf{Z}) \to H_m(V, \mathbf{Z})$. Now (2.13) follows by taking $W = L(S)$, $Y = P$ in:

Lemma (2.14) *Let V be a smooth projective variety defined over $\mathbf{C}$, $\dim V = d + 1$, $\phi\colon W \to V$ a proper morphism, generically finite of degree k (with W irreducible), $X \hookrightarrow V$ a smooth hyperplane section, and $Y = \phi^{-1}(X)$. Then the image of $\phi_*\colon H_d(Y, \mathbf{Z}) \to H_d(X, \mathbf{Z})$ contains the vanishing cycles.*

Proof

Step a. We may assume X generic. Indeed, let P^* denote the parameter space for hyperplanes $X \hookrightarrow V$. We have families

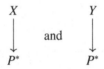

with fibres, respectively, hyperplanes $X_t \subset V$ and inverse images $Y_t = \phi^{-1}(X_t) \subset W$, $t \in P^*$. For $t_0 \in P^*$, there exists a neighborhood $t_0 \in U \subset P^*$ such that X_U and Y_U contract onto X_{t_0} and Y_{t_0} respectively. For $t \in U$ we get a commutative diagram of specializations

$$
\begin{array}{ccc}
H_d(Y_t, \mathbf{Z}) & \longrightarrow & H_d(Y_{t_0}, \mathbf{Z}) \\
\phi_* \downarrow & & \downarrow \phi_* \\
H_d(X_t, \mathbf{Z}) & \longrightarrow & H_d(X_{t_0}, \mathbf{Z}).
\end{array}
$$

Assuming X_t, X_{t_0} smooth, the bottom horizontal arrow is an isomorphism. Thus to prove the right-hand vertical arrow surjective, it suffices to show the left-hand one is.

Step b. The sheaf $R^d \pi_* \mathbf{Z}$ is constructible on P^*, where $\pi\colon Y \to P^*$. Let $U \subset P^*$ be a non-empty Zariski open set such that $R^d \pi_* \mathbf{Z}$ is locally constant on U. Let $\ell \in P^*$ be a general line corresponding to a Lefschetz pencil $\{X_t\}$ on V.

Let $X_0 = X$ and let $X_{t_1}, \ldots, X_{t_n}$ denote the singular fibres. Choose paths $\tau_i\colon [0, 1] \to \ell$ such that $\tau_i(0) = 0$ and $\tau_i(1) = t_i$. We may suppose $\tau_i(x) \in U$ for $0 \leq x \leq 1$. To each τ_i there corresponds a vanishing cycle $\delta_i \in H_d(X, \mathbf{Z})$, and these vanishing cycles generate the group of vanishing cycles. Let us show, for example, that $\delta_1 \in \text{Image}(\phi_*)$.

Since $\{X_t\}$ is general, we may assume that the singular points of the fibres,

$p_1, \ldots, p_n \in V$ lie in the open set over which ϕ is étale. Let N be a neighborhood of p_1 in V such that $\phi^{-1}(N) = M^{(1)} \cup \cdots \cup M^{(k)}$, where $\phi : M^{(i)} \xrightarrow{\sim} N$ for each i. Let ε be close to 1, and let $\phi_\varepsilon : Y_\varepsilon \to X_\varepsilon$ be the situation over the point $\tau_1(\varepsilon)$. Since $|\varepsilon - 1|$ is small, the vanishing cycle δ_ε is "close to disappearing" on X_ε – that is, it is supported on $X_\varepsilon \cap N$ – so it can be lifted to $Y_\varepsilon \cap M^{(i)} \to X_\varepsilon \cap N$. But now the fact that $\tau_1([0, \varepsilon]) \subseteq U$ gives

$$
\begin{array}{ccc}
H_d(Y_0) & \xrightarrow{\;\approx\;} & H_d(Y_\varepsilon) \\
\phi_{0*} \downarrow & & \downarrow \phi_{\varepsilon*} \\
H_d(X_0) & \xrightarrow{\;\approx\;} & H_d(X_\varepsilon)
\end{array}
$$

$$
\delta_1 \longmapsto \delta_\varepsilon.
$$

Since $\delta_\varepsilon \in \mathrm{Image}(\phi_{\varepsilon*})$, we get $\delta_1 \in \mathrm{Image}(\phi_{0*})$. This completes the proof of Lemma (2.14). □

Remarks on the literature

Intermediate jacobians were first introduced by Weil, and then modified to vary holomorphically in a family by Griffiths. The cycle map is defined in Griffiths [6] and the first application to Fano varieties, the proof of the irrationality of the cubic threefold, is in Clemens and Griffiths [4]. The first investigation of the relation between A^2 and J^2 for Fano threefolds is in Murre [8]. I have also borrowed heavily from Tyurin [10], particularly regarding (2.9) and the proof of (2.12).

References for Lecture 2

[1] A. Beauville, Variétés de Prym et jacobiennes intermédiaires, *Ann. Sci. Ecole Norm. Sup. (4)*, **10** (1977), 304–391.

[2] S. Bloch, An example in the theory of algebraic cycles, pp. 1–29 in *Algebraic K-Theory*, Lecture Notes in Math., no. 551, Springer, Berlin (1976).

[3] S. Bloch and J. P. Murre, On the Chow groups of certain types of Fano threefolds, *Compositio Math.*, **39** (1979), 47–105.

[4] C. H. Clemens and P. A. Griffiths, The intermediate jacobian of the cubic threefold, *Ann. of Math. (2)*, **95** (1972), 281–356.

[5] P. Deligne, Théorie de Hodge. I, pp. 425–430 in *Actes du Congrès International des Mathématiciens (Nice, 1970)*, vol. 1, Gauthier-Villars, Paris (1971).

[6] P. Griffiths, On the periods of certain rational integrals. I, II, *Ann. of Math. (2)*, **90** (1969), 460–495; **90** (1969), 496–541.

[7] P. Griffiths and J. Harris, *Principles of Algebraic Geometry*, Wiley, New York (1978). [Reprinted 1994.]

[8] J. P. Murre, Algebraic equivalence modulo rational equivalence on a cubic threefold, *Compositio Math.*, **25** (1972), 161–206.

[9] B. R. Tennison, On the quartic threefold, *Proc. London Math. Soc. (3)*, **29** (1974), 714–734.

[10] A. N. Tyurin, Five lectures on three dimensional varieties, *Uspehi Mat. Nauk*, **27** (1972), no. 5 (167), 3–50. [Translation: Russian Math. Surveys, **27** (1972), no. 5, 1–53.]

[11] A. Wallace, *Homology Theory on Algebraic Varieties*, Pergamon Press, New York (1958).

[12] D. I. Lieberman, Higher Picard varieties, *Amer. J. Math.*, **90** (1968), 1165–1199.

3

Curves on threefolds and intermediate jacobians – the relative case

Let X be a smooth quasi-projective variety over a field k, and $Y \subset X$ a divisor with normal crossings. A problem of considerable importance is to develop a theory of *relative* algebraic cycles for X relative to Y. This problem arises geometrically – for example, if X is the resolution of a singular variety X' and Y is the exceptional locus, and one wishes to study cycles on X' (compare the theory of generalized jacobians of Rosenlicht [4] and Serre [5]). It also arises in K-theory. Indeed it seems likely that the higher K-groups of a variety admit a filtration whose successive quotients can be interpreted up to torsion as relative cycle groups in much the same way that the γ-filtration on K_0 relates K_0 to the usual Chow group. (This is a point I hope to take up elsewhere.) Finally, and rather surprisingly, there are compelling arithmetic reasons to develop such a theory. An interesting example will be treated at length in Lectures 8 and 9.

One feature of such a theory when $k \subset \mathbf{C}$ is a theory of cycle classes for relative codimension-r cycles in the intermediate jacobians $J^r(X, Y)$ associated to the relative cohomology group $H^{2r-1}(X, Y)$ (see below). $J^r(X, Y)$ will be a non-compact complex torus, so one may quotient by the maximal compact real subtorus to obtain cycle classes in a real or complex vector space. In some circumstances, one expects these cycle classes to fill out a lattice whose volume is linked to some value of a zeta function in much the same way the classical regulator is associated to the value at 1 of the zeta function of a number field. Indeed, the classical case is precisely the case $X = \mathbf{P}^1$ and $Y = \{0, \infty\}$. The relative cycles are points in $\mathbf{P}^1 - \{0, \infty\} = \mathbf{G}_m$ corresponding to units in a given number field.

In this lecture we will work out in some detail the construction of relative cycle classes of codimension 2 in the intermediate jacobian $J^2(X, Y)$ for the particular case $X = C \times \mathbf{P}^1 \times \mathbf{P}^1$ with C a nonsingular projective curve and $Y = (C \times \mathbf{P}^1 \times \{0, \infty\}) \cup (C \times \{0, \infty\} \times \mathbf{P}^1) := C \times \#$.

A one-dimensional algebraic cycle $\gamma = \sum n_i \gamma_i$ on $C \times \mathbf{P}^1 \times \mathbf{P}^1$ will be called a *relative* cycle if

(i) all the γ_i meet $C \times \#$ properly, and
(ii) if $\ell \subset \#$ is any of the four lines, $\gamma \cdot C \times \ell = 0$ as a cycle.

These two conditions are sufficient if no component of γ meets any of the "corners" $C \times \{i, j\}$, for $i, j = 0, \infty$. However, for the examples it is convenient to permit γ to meet the corners, but only in a rather special way. Let $p \in C \times \#$, $p = (p_0, p_1, p_2)$, and let $M_\varepsilon = D_0 \times D_{1,\varepsilon} \times D_{2,\varepsilon}$ be a product of closed disks centered at the p_i where D_0 is fixed and D_1 and D_2 have radius $\varepsilon \ll 1$, so $p \in M_\varepsilon \subset C \times \mathbf{P}^1 \times \mathbf{P}^1$. To fix the ideas we will consider only the hard case where $p_1, p_2 = 0, \infty$. The discussion when p is a smooth point on $C \times \#$ is analogous but easier and is left for the reader. Assume $p \in \text{Supp } \gamma$. Condition (ii) above implies for ε small that $\gamma \cap M_\varepsilon$ is trivial in $H_2(M_\varepsilon, \partial M_\varepsilon) \cong H_1(\partial M_\varepsilon)$. Indeed, we can assume no component of γ meets $\partial D_0 \times D_{1,\varepsilon} \times D_{2,\varepsilon}$, so

$$\partial(\gamma \cap M_\varepsilon) \simeq n_1 p_0 \times \partial D_1 \times p_2 + n_2 p_0 \times p_1 \times \partial D_2,$$

where n_i is the multiplicity of intersection of γ with $C \times \ell_i$ ($\ell_i = $ corresponding line) at p. By hypothesis both $n_i = 0$.

Now fix an ε_0 for which the above discussion is valid and let Γ be a 3-chain on M_{ε_0} with $\partial \Gamma = M_{\varepsilon_0} \cap \gamma + \tau_{\varepsilon_0}$, where τ_{ε_0} is supported on $D_0 \times \partial D_{1,\varepsilon_0} \times D_{2,\varepsilon_0} \cup D_0 \times D_{1,\varepsilon_0} \times \partial D_{2,\varepsilon_0}$. Define Γ_ε on M_ε for $\varepsilon \leq \varepsilon_0$ by excision. For γ to be a relative cycle we require the existence of Γ_{ε_0} such that $\Gamma_{\varepsilon_0} \cap C \times \# = \{p\}$ and such that, moreover, writing x_1, x_2 for the two obvious meromorphic functions on $\mathbf{P}^1 \times \mathbf{P}^1$, we have

$$(3.1) \qquad \lim_{\varepsilon \to 0} \int_{\tau_\varepsilon} \frac{dx_1}{x_1} \wedge \frac{dx_2}{x_2} = 0.$$

(The reader should check this is always the case when p is a smooth point of $C \times \#$.)

Example (3.2) Suppose $p = (0, 0, 0)$, and that γ has two irreducible components through p, one with coefficient $+1$ and the other with -1, parameterized locally by

$$\{(x, a_1 x^{r_1} + \text{hot}, b_1 x^{s_1} + \text{hot})\}$$
$$\{(x, a_2 x^{r_2} + \text{hot}, b_2 x^{s_2} + \text{hot})\} \qquad (\text{hot} = \text{higher order terms}).$$

Changing coordinates (the first coordinate is immaterial) we can rewrite these parameterizations

$$\{(-, x, b_1 a_1^{-s_1/r_1} x^{s_1/r_1} + \text{hot})\},$$
$$\{(-, x, b_2 a_2^{-s_2/r_2} x^{s_2/r_2} + \text{hot})\}.$$

On $D_0 \times \partial D_{1,\varepsilon} \times D_{2,\varepsilon}$, a natural chain to take is

$$\tau'_\varepsilon = \left\{ (\lambda(-) + (1-\lambda)(-), x, \lambda(b_1 a_1^{-s_1/r_1} x^{s_1/r_1} + \text{hot}) \right.$$
$$\left. + (1-\lambda)(b_2 a_2^{-s_2/r_2} x^{s_2/r_2} + \text{hot}) \,\middle|\, |x| = \varepsilon, 0 \le \lambda \le 1 \right\}.$$

Assume now $s_1/r_1 = s_2/r_2$. One gets formally

$$\int_{\tau'_\varepsilon} \frac{dx_1}{x_1} \wedge \frac{dx_2}{x_2} = \frac{2\pi i}{r_1} \log\left(\frac{b_1^{r_1} a_2^{s_1}}{b_2^{r_1} a_1^{s_1}} \right) + o(1).$$

A similar analysis is valid on $D_0 \times D_{1,\varepsilon} \times \partial D_{2,\varepsilon}$, and we find the conditions

(3.2.1) $$\frac{r_1}{s_1} = \frac{r_2}{s_2}, \qquad b_1^{r_1} a_2^{s_1} = b_2^{r_1} a_1^{s_1}$$

lead to a relative cycle in (3.2). These conditions will really only begin to make sense in Lecture 8 after we have studied K-theory and the tame symbol.

In Lecture 2, we defined the cycle map by integration over 3-chains. The process will be similar here, except that we will deal with a family of 3-chains Δ_ε and with a limit of integrals over Δ_ε, as $\varepsilon \to 0$. Let $\gamma = \sum n_i \gamma_i$ be a relative algebraic 1-cycle on $C \times \mathbf{P}^1 \times \mathbf{P}^1$. Let $p(1), \ldots, p(N)$ be the points where components of γ meet $C \times \#$, and let $M_\varepsilon(j)$ be a neighborhood of $p(j)$ as above, $\varepsilon \le \varepsilon_0$. Taking these neighborhoods to be sufficiently small we may assume the disjoint union $M_\varepsilon = \coprod M_\varepsilon(j) \subset C \times \mathbf{P}^1 \times \mathbf{P}^1$. Let $M_\varepsilon^0 \subset M_\varepsilon$ be the interior.

On each neighborhood $M_{\varepsilon_0}(j)$ we can define a 3-chain $\Gamma_{\varepsilon_0}(j)$ as before with $\partial \Gamma_{\varepsilon_0}(j) = \gamma \cap M_{\varepsilon_0}(j) - \tau_{\varepsilon_0}(j)$ with $\tau_{\varepsilon_0}(j)$ supported on $\partial M_{\varepsilon_0}(j)$. We define $\Gamma_\varepsilon(j)$ for $\varepsilon \le \varepsilon_0$ by excision, so $\partial \Gamma_\varepsilon(j) = \gamma \cap M_\varepsilon(j) - \tau_\varepsilon(j)$, where $|\tau_\varepsilon(j)| \subset \partial M_\varepsilon(j)$. We take $\Gamma_\varepsilon = \coprod \Gamma_\varepsilon(j)$, $\tau_\varepsilon = \coprod \tau_\varepsilon(j)$.

Let γ_ε be the homology 2-chain on $C \times \mathbf{G}_m \times \mathbf{G}_m - M_\varepsilon^0$ (here γ_ε is a 2-cycle relative to ∂M_ε) obtained from γ by excision. The cycle $\gamma_\varepsilon - \tau_\varepsilon$ is a 2-cycle on $C \times \mathbf{G}_m \times \mathbf{G}_m - M_\varepsilon^0$. Let $D = (\mathbf{G}_m \times \{1\}) \cup (\{1\} \times \mathbf{G}_m)$.

Lemma (3.3) *There exists a 3-chain Δ_{ε_0} on $C \times \mathbf{G}_m \times \mathbf{G}_m - M_{\varepsilon_0}^0$ with $\partial \Delta_{\varepsilon_0} = \gamma_{\varepsilon_0} - \tau_{\varepsilon_0} + E_{\varepsilon_0}$, where E_{ε_0} is supported on $C \times D$.*

Proof The inclusion $C \times \mathbf{G}_m \times \mathbf{G}_m - M_{\varepsilon_0}^0 \to C \times \mathbf{G}_m \times \mathbf{G}_m$ is easily seen to be a homotopy equivalence. The Künneth decomposition gives therefore

$$H_2(C \times \mathbf{G}_m \times \mathbf{G}_m - M_{\varepsilon_0}^0) \cong H_2(C) \otimes H_1(C) \otimes H_1(C) \otimes \mathbf{Z}.$$

Note that

$$\frac{dx_1}{x_1} \wedge \frac{dx_2}{x_2} \bigg|_{\gamma_\varepsilon} = 0$$

by reason of type, so

$$\int_{\gamma_\varepsilon - \tau_\varepsilon} \frac{dx_1}{x_1} \wedge \frac{dx_2}{x_2} = -\int_{\tau_\varepsilon} \frac{dx_1}{x_1} \wedge \frac{dx_2}{x_2} \to 0, \quad \text{as } \varepsilon \to 0.$$

This implies that the class of $\gamma_{\varepsilon_0} - \tau_{\varepsilon_0}$ lies in $H_2(C) \oplus H_1(C) \oplus H_1(C)$. Since this subgroup is the image of $H_2(C \times D)$, we can find Δ_{ε_0} as claimed.   □

We define Δ_ε for $\varepsilon \leq \varepsilon_0$ by considering the chain $\Delta_{\varepsilon_0} - \Gamma_{\varepsilon_0}$ and excising away M_ε^0. We have

$$(3.3.1) \qquad \partial\Delta_\varepsilon = \gamma_\varepsilon - \tau_\varepsilon + E_\varepsilon, \qquad \text{Supp } E_\varepsilon \subset C \times D.$$

The next step is to define the appropriate intermediate jacobian.

Definition (3.4) (Deligne [2])   A *mixed Hodge structure* consists of

(a) a free **Z**-module of finite type $H_{\mathbf{Z}}$,
(b) a finite increasing filtration (*weight filtration*) W on $H_{\mathbf{Q}} = H_{\mathbf{Z}} \otimes_{\mathbf{Z}} \mathbf{Q}$,
(c) a finite decreasing filtration F^* on $H_{\mathbf{C}}$ (*Hodge filtration*).

These data are required to satisfy the condition that there should exist on $\mathrm{gr}_W H_{\mathbf{C}}$ a (necessarily unique) bigradation by subspaces H^{pq} such that

(i) $\mathrm{gr}_W^n H_{\mathbf{C}} = \oplus_{p+q=n} H^{pq}$,
(ii) the filtration F^* induces on $\mathrm{gr}_W H_{\mathbf{C}}$ the filtration

$$\mathrm{gr}_W(F)^p = \oplus_{p' \geq p} H^{p',q}$$

(iii) $\overline{H}^{pq} = H^{qp}$.

Suppose now we are given a mixed Hodge structure H with weights $\leq 2r-1$, that is $\mathrm{gr}_W^m H_{\mathbf{Q}} = 0$, for $m > 2r - 1$. The same prescription as before, namely

$$J^r = H_{\mathbf{C}}/F^r + H_{\mathbf{Z}},$$

will in this case give us an abelian complex Lie group; that is, the image of $H_{\mathbf{Z}}$ in $H_{\mathbf{C}}/F^r$ is discrete but not necessarily cocompact.

Example (3.5)   Let X be a smooth projective variety, and $Y \subset X$ a closed (not necessarily smooth or even connected) subvariety. In general, for a proper variety Y, Deligne has defined [1] a mixed Hodge structure with weights $\leq k$ on $H^k(Y, \mathbf{C})$. He also defined [1] a Hodge structure on the relative cohomology $H^k(X, Y; \mathbf{C})$ of weight $\leq k$ compatible with the morphisms $H^{k-1}(Y, \mathbf{C}) \to H^k(X, Y; \mathbf{C}) \to H^k(X, \mathbf{C})$ in the relative cohomology sequence. We can thus consider the relative intermediate jacobian

$$J^r(X, Y) := H^{2r-1}(X, Y; \mathbf{C})/F^r + H^{2r-1}(X, Y; \mathbf{Z}).$$

More specific example (3.6) Take $X = C \times \mathbf{P}^1 \times \mathbf{P}^1$, $Y = C \times \#$ as above, and let $r = 2$. Actually, in this case the relative cohomology has a Künneth decomposition, and we will focus on $H^1(C) \otimes H^2(\mathbf{P}^1 \times \mathbf{P}^1, \#)$. Combining the exact sequence of relative cohomology for $(\mathbf{P}^1 \times \mathbf{P}^1, \#)$ with the Mayer–Vietoris sequence

$$\cdots \to \oplus_4 H^{i-1}(\mathbf{P}^1) \to \oplus_4 H^{i-1}(\mathrm{pt}) \to H^i(\#) \to H^i(\mathbf{P}^1 \times \mathbf{P}^1) \to \cdots,$$

one checks that there is an isomorphism of Hodge structures

$$H^2(\mathbf{P}^1 \times \mathbf{P}^1, \#) \cong H^0(\mathrm{pt}).$$

Thus

$$F^2(H^1(C) \otimes H^2(\mathbf{P}^1 \times \mathbf{P}^1, \#)) = F^2 H^1(C) \otimes H^2(\mathbf{P}^1 \times \mathbf{P}^1, \#) = (0),$$

so the intermediate jacobian associated to this piece of $H^3(C \times \mathbf{P}^1 \times \mathbf{P}^1, C \times \#)$ is $H^1(C, \mathbf{C})/H^1(C, \mathbf{Z}) \cong H^1(C, \mathbf{C}^*)$.

We next define our cycle map

$$\left\{ \begin{array}{c} \text{relative algebraic} \\ \text{1-cycles on } C \times \mathbf{P}^1 \times \mathbf{P}^1 \end{array} \right\} \to H^1(C, \mathbf{C}^*).$$

Given a relative cycle γ we choose a family of 3-chains Δ_ε depending on ε such that, with notation as before,

$$\partial \Delta_\varepsilon = \gamma_\varepsilon - \tau_\varepsilon + E_\varepsilon, \qquad \mathrm{Supp}\, E_\varepsilon \subset C \times D.$$

If η is a closed global C^∞ 1-form on C, we define

$$(3.7) \qquad P_\gamma(\eta) = \lim_{\varepsilon \to 0} \frac{-1}{4\pi^2} \int_{\Delta_\varepsilon} \eta \wedge \frac{dx_1}{x_1} \wedge \frac{dx_2}{x_2}.$$

Lemma (3.7.1)

(i) *The limit in (3.7) is defined and* $< \infty$.

(ii) $P_\gamma(df) = 0$ *if* f *is a* C^∞ *function on* C.

(iii) *Different choices of* Δ_ε *change* P_γ *by an integral period of* C; *that is,* $P_\gamma(\eta)$ *changes by* $\int_\alpha \eta$ *for some integral 1-cycle* α *on* C *independent of* η.

Proof (i) It suffices to examine the situation in a neighborhood M of the finite set of points where $\mathrm{Supp}\, \gamma$ meets $C \times \#$. On M we can write $\eta = df$, and Stokes' theorem gives

$$\int_{\Delta_\varepsilon \cap M} df \wedge \frac{dx_1}{x_1} \wedge \frac{dx_2}{x_2} = \int_{-\tau_\varepsilon} f \frac{dx_1}{x_1} \wedge \frac{dx_2}{x_2}.$$

$$\left(\text{Note } \frac{dx_1}{x_1} \wedge \frac{dx_2}{x_2} \bigg|_{\gamma_\varepsilon} = \frac{dx_1}{x_1} \wedge \frac{dx_2}{x_2} \bigg|_D = 0. \right)$$

Given our hypotheses about relative cycles, it is easy to show that the right-hand side tends to 0 with ε, which is enough to deduce existence of the limit.

(ii) The argument is precisely the same as for (i), except the integral over $\Delta_\varepsilon \cap M$ is replaced by the integral over Δ_ε.

(iii) Suppose Δ'_ε is another family of 3-chains such that

$$\partial \Delta'_\varepsilon = \gamma_\varepsilon - \tau'_\varepsilon + E'_\varepsilon, \qquad E'_\varepsilon \subset C \times D.$$

There will exist a 3-chain T_ε supported on the boundary of an M_ε (notation as (3.2)) such that $\partial T_\varepsilon = \tau_\varepsilon - \tau'_\varepsilon$. Since $H_2(C \times D) \hookrightarrow H_2(C \times \mathbf{G}_m \times \mathbf{G}_m)$ there will exist a 3-chain U_ε supported on $C \times D$ such that $\partial U_\varepsilon = E_\varepsilon - E'_\varepsilon$. Then $\Delta_\varepsilon - \Delta'_\varepsilon - U_\varepsilon + T_\varepsilon$ is a 3-cycle. Since $\frac{1}{2\pi i} \frac{dx_j}{x_j}$ represents an integral cohomology class on $\mathbf{G}_m$, it follows that

$$\frac{-1}{4\pi^2} \int_{\Delta_\varepsilon - \Delta'_\varepsilon - U_\varepsilon + T_\varepsilon} \eta \wedge \frac{dx_1}{x_1} \wedge \frac{dx_2}{x_2} = \text{period of } \eta \text{ on } C.$$

Clearly $\int_{U_\varepsilon} \eta \wedge \frac{dx_1}{x_1} \wedge \frac{dx_2}{x_2} = 0$. Moreover, for ε small, $\eta = df$ in some neighborhood of M_ε, so

$$\int_{T_\varepsilon} \eta \wedge \frac{dx_1}{x_1} \wedge \frac{dx_2}{x_2} = \int_{\tau_\varepsilon - \tau'_\varepsilon} f \frac{dx_1}{x_1} \wedge \frac{dx_2}{x_2} \to 0,$$

as in (i). We conclude

$$\lim_{\varepsilon \to 0} \frac{-1}{4\pi^2} \int_{\Delta_\varepsilon - \Delta'_\varepsilon} \eta \wedge \frac{dx_1}{x_1} \wedge \frac{dx_2}{x_2} = \text{period of } \eta. \qquad \square$$

We now have our cycle map

$$\left\{ \begin{matrix} \text{algebraic 1-cycles on } C \times \mathbf{P}^1 \times \mathbf{P}^1 \\ \text{relative to } C \times \# \end{matrix} \right\} \begin{matrix} \longrightarrow \\ \gamma \mapsto P_\gamma \end{matrix} H^1(C, \mathbf{C}^*).$$

Lecture 8 will be devoted to a special example, where the period will be related to the Hasse–Weil zeta function of the curve C. For this purpose, it will be important to *factor the non-compact torus* $H^1(C, \mathbf{C}^*)$ *by its maximal compact subgroup* – that is, to consider for η a real closed 1-form on C

$$(3.8) \qquad \text{Im } P_\gamma(\eta) = \lim_{\varepsilon \to 0} \text{Im} \left(\frac{-1}{4\pi^2} \int_{\Delta_\varepsilon} \eta \wedge \frac{dx_1}{x_1} \wedge \frac{dx_2}{x_2} \right).$$

One virtue of this is that the assignment $\gamma \mapsto \operatorname{Im} P_\gamma$ extends naturally to all (not necessarily relative) algebraic 1-cycles on $C \times \mathbf{P}^1 \times \mathbf{P}^1$ which meet $C \times \#$ properly:

$$\operatorname{Im} P : \left\{ \begin{array}{c} \text{all algebraic 1-cycles on } C \times \mathbf{P}^1 \times \mathbf{P}^1 \\ \text{meeting } C \times \# \text{ properly} \end{array} \right\} \to H^1(C, \mathbf{R}).$$

In order to see this, let γ now denote any such algebraic 1-cycle on $C \times \mathbf{P}^1 \times \mathbf{P}^1$, and let $V_\varepsilon = \coprod_k V_{k,\varepsilon} \subset C$ be a union of small disjoint disks of radius ε about the C coordinates of points in $|\gamma| \cap C \times \#$. Then γ gives a class by excision in $H_2(C_\varepsilon \times \mathbf{G}_m \times \mathbf{G}_m, \partial V_\varepsilon \times \mathbf{G}_m \times \mathbf{G}_m)$, where $C_\varepsilon = C - V_\varepsilon^0$.

Writing $\gamma'_\varepsilon = \gamma - V_\varepsilon^0 \times \mathbf{P}^1 \times \mathbf{P}^1$, an analysis of the homology group $H_2(C_\varepsilon \times \mathbf{G}_m \times \mathbf{G}_m, \partial V_\varepsilon \times \mathbf{G}_m \times \mathbf{G}_m)$ shows that there exists a 3-chain Δ_ε on $C_\varepsilon \times \mathbf{G}_m \times \mathbf{G}_m$ with $\partial \Delta_\varepsilon = \gamma'_\varepsilon + E_\varepsilon + F_\varepsilon$, where $|E_\varepsilon| \subset (C_\varepsilon \times D)$ and $|F_\varepsilon| \subset \partial V_\varepsilon \times \mathbf{G}_m \times \mathbf{G}_m$. As before, we can construct a continuous family of Δ_ε depending on ε as $\varepsilon \to 0$.

Lemma (3.9) *Let η be a closed real 1-form on C. Then*

$$\operatorname{Im} \lim_{\varepsilon \to 0} \int_{\Delta_\varepsilon} \eta \wedge \frac{dx_1}{x_1} \wedge \frac{dx_2}{x_2}$$

is independent of the choice of Δ_ε.

Proof Let $W_\varepsilon \subset C$ be the connected, simply connected set obtained by linearly ordering the disks $V_{k,\varepsilon}$ and connecting them by narrow strips.

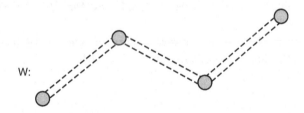

W:

Since W is simply connected, we have $\eta|_W = df$.

Suppose now $\bar{\Delta}_\varepsilon$ is another 3-chain with

$$\partial \bar{\Delta}_\varepsilon = \gamma'_\varepsilon + \bar{E}_\varepsilon + \bar{F}_\varepsilon,$$

$$|\bar{E}_\varepsilon| \subset C \times D, \qquad |\bar{F}_\varepsilon| \subset \partial V_\varepsilon \times \mathbf{G}_m \times \mathbf{G}_m.$$

Let $S = (W \times \mathbf{G}_m \times \mathbf{G}_m) \cup (C \times D)$. One checks easily that $H_2(S) \to H_2(C \times \mathbf{G}_m \times \mathbf{G}_m)$. Since $E_\varepsilon + F_\varepsilon - \bar{E}_\varepsilon - \bar{F}_\varepsilon$ is a 2-cycle on S which bounds on $C \times \mathbf{G}_m \times \mathbf{G}_m$, there exist 3-chains T on $W \times \mathbf{G}_m \times \mathbf{G}_m$ and U on $C \times D$ such that $\partial T + \partial U = \partial(\Delta_\varepsilon - \bar{\Delta}_\varepsilon)$. In particular $\Delta_\varepsilon - \bar{\Delta}_\varepsilon - T - U$ is an integral 3-cycle, and

$$\operatorname{Im} \int_{\Delta_\varepsilon - \bar{\Delta}_\varepsilon} \eta \wedge \frac{dx_1}{x_1} \wedge \frac{dx_2}{x_2} = \operatorname{Im} \int_{T+U} = \operatorname{Im} \int_T = \operatorname{Im} \int_{\partial T} f \frac{dx_1}{x_1} \wedge \frac{dx_2}{x_2}.$$

Note that $\frac{dx_1}{x_1} \wedge \frac{dx_2}{x_2}$ actually represents a class in $H^2(\mathbf{G}_m \times \mathbf{G}_m, D; \mathbf{R})$. Also $\partial T = \sum_k F_{\varepsilon,k} - \bar{F}_{\varepsilon,k}$ rel $(C \times D)$. Thus $F_{\varepsilon,k} - \bar{F}_{\varepsilon,k}$ is a relative 2-cycle on $V_{k,\varepsilon} \times \mathbf{G}_m \times \mathbf{G}_m$, where say $V_{k,\varepsilon}$ is a closed ε-disk about a point $t_k \in C$. It will suffice therefore to show

$$\lim_{\varepsilon \to 0} \operatorname{Im} \int_{F_{k,\varepsilon} - \bar{F}_{k,\varepsilon}} (f - f(t_k)) \frac{dx_1}{x_1} \wedge \frac{dx_2}{x_2} = 0.$$

This is straightforward. □

References for Lecture 3

[1] P. Deligne, Théorie de Hodge III, *Inst. Hautes Etudes Sci. Publ. Math.*, no. 44 (1974), 5–77.

[2] P. Deligne, Poids dans la cohomologie des variétés algébriques, pp. 79–85 in *Proceedings of the International Congress of Mathematicians (Vancouver, B.C., 1974)*, vol. 1 (1975).

[3] P. Griffiths, On the periods of certain rational integrals. I, II, *Ann. of Math. (2)*, **90** (1969), 460–495; **90** (1969), 496–541.

[4] M. Rosenlicht, Generalized Jacobian varieties, *Ann. of Math.*, **59** (1954), 505–530.

[5] J. P. Serre, *Groupes algébriques et corps de classes*, Hermann, Paris (1959). [Second edition, 1975; reprinted, 1984.]

4

K-theoretic and cohomological methods

In this lecture we will discuss briefly some of Quillen's ideas about K-theory, and their extensions to various other "cohomology theories". Basic references are Quillen [3] and Bloch and Ogus [6]. The main results are the cohomological interpretations of the cycle groups

$$H^p(X, \mathcal{K}_p) \cong CH^p(X),$$

$$H^p(X, \mathcal{H}^p(\mathbf{Z})) \cong CH^p(X)/A^p(X) \qquad (X \text{ defined over } \mathbf{C}),$$

$$H^p(X, \mathcal{H}^p(\mu_\ell^{\otimes p})) \cong CH^p(X)/\ell CH^p(X).$$

(For notations, see below.)

Let C be a small category. The *nerve* of C, NC is the simplicial set with p-simplices the set of diagrams

$$x_0 \rightarrow x_1 \rightarrow \cdots \rightarrow x_p \quad \text{with } x_i \in \mathrm{Ob}\, C, \text{ arrows in Morph } C.$$

Face operators arise by omitting objects and composing arrows. Degeneracies are defined by inserting identity maps. The classifying space for NC is denoted BC.

An *exact category* is an additive category $\mathcal{M}$ which can be embedded as a full subcategory of an abelian category $\mathcal{A}$ in such a way that if

$$0 \rightarrow A' \rightarrow A \rightarrow A'' \rightarrow 0$$

is exact in $\mathcal{A}$ and A', A'' are isomorphic to objects in $\mathcal{M}$, then A is isomorphic to an object of $\mathcal{M}$. (For an intrinsic characterization, see Quillen [3]). Given a diagram in

$$0 \rightarrow M' \xrightarrow{i} M \xrightarrow{j} M'' \rightarrow 0$$

which is exact in $\mathcal{A}$, Quillen refers to i as an admissible monomorphism and

j as an admissible epimorphism. A functor $F: M \to M'$ on exact categories is exact if it preserves exact sequences.

The exact category M has a zero object, from which it follows that BM is contractible (exercise). Quillen's idea is to build from M a new category QM in such a way that

$$\pi_1(BQM) \cong k_0(M)$$

$$:= \frac{\mathbf{Z}^{\mathrm{Ob}\,M}}{\langle [n] - [n'] - [n''] \mid 0 \to M' \to M \to M'' \to 0 \text{ exact in } M \rangle}.$$

He then defines

$$K_n(M) := \pi_{n+1}(BQM).$$

Objects in QM will be the same as objects in M, but a morphism in QM, $M_1 \xrightarrow[QM]{} M_2$, is by definition an isomorphism in M of M_1 with a subquotient of M_2, that is a filtration $M' \subset M'' \subset M_1$ with M_2/M'', M_2/M', and M''/M' in M and an isomorphism $\Theta: M_1 \cong M''/M'$. In other words, a morphism is a diagram in M

$$M_1 \xleftarrow{\ j\ } M'' \xhookrightarrow{\ i\ } M_2,$$

where i and j are admissible mono and epimorphisms respectively. For example, for any $M \in \mathrm{Ob}\,M$, there are two canonical maps $0 \to M$ in QM given by the obvious sub and quotient arrows in M:

$$0 \rightrightarrows M.$$

One defines in this way a loop $0 \rightrightarrows M$, and the map

$$K_0(M) \to \pi_1(BQM)$$

sends $[M] \mapsto 0 \rightrightarrows M$. (A direct proof that this is an isomorphism, or even a well-defined map might be awkward. The reader is advised to consult Quillen [3], where a preliminary discussion of covering spaces for spaces BC helps to grease the skids.) Note finally that an exact functor $F: M \to M'$ induces a functor $QM \to QM'$.

When M is the category of finitely generated projectives over a ring A (associated with 1), Quillen gave another definition of $K_*(M) := K_*(A)$. He defined an H-space $BGL(A)^+$ and a map from the classifying space of the infinite general linear group $(:= \varinjlim BGL_n(A))$

$$BGL(A) \to BGL(A)^+,$$

which was acyclic (i.e. induced an isomorphism on homology with coefficients in any local system) and hence (by obstruction theory) was universal for maps $BGL(A) \to X$, X any H-space. He defined $K_n(A) = \pi_n(BGL(A)^+)$, $n \geq 1$, and proved $\pi_n(BGL(A)^+) \cong \pi_{n+1}(B Q M)$ (Quillen and Grayson [4]). It is not hard to show from this

$$K_1(A) \cong GL(A)/[GL(A), GL(A)] \cong GL(A)/E(A),$$
$$K_2(A) \cong H_2(E(A), \mathbf{Z}).$$

Here $E(A)$ is the group of *elementary* matrices. One knows [1] that $E(A) = [GL(A), GL(A)]$ and $[E(A), E(A)] = E(A)$. These definitions agreed, therefore, with definitions given earlier by Bass [1] and Milnor [2] for K_1 and K_2. K_2 can also be defined to be the center of the universal central extension of $E(A)$:

$$0 \to K_2(A) \to \mathrm{St}(A) \to E(A) \to 1.$$

Here $\mathrm{St}(A)$ denotes the Steinberg group [2], and universal means that any central extension of $E(A)$ by an abelian group G arises via pushout from a unique homomorphism $K_2(A) \to G$.

Here is a compendium of useful facts about K_0, K_1, K_2 for a commutative *local* ring R:

(i) Projective modules over R are free, so $K_0(R) \cong \mathbf{Z}$.
(ii) $E(R) = SL(R)$, so $K_1(R) \cong R^*$, the unit group of R.
(iii) There is a bilinear pairing

$$R^* \otimes_{\mathbf{Z}} R^* \to K_2(R),$$

sending $r_1 \otimes r_2$ to the Steinberg symbol $\{r_1, r_2\}$. This pairing satisfies the relations

(a) $\{r, -r\} = 1$, $\quad r \in R^*$,
(b) $\{r, 1 - r\} = 1$ (Steinberg relation), $\quad r, 1 - r \in R^*$,
(c) $\{r, s\}\{s, r\} = 1$, $\quad r, s \in R^*$.

$K_2(R)$ is generated by Steinberg symbols, and when R is a field, the above list of relations is complete (indeed redundant, as (a) and (c) follow from (b)).

We now merely list some of the main results proved by Quillen about the spaces $B Q M$ for general exact categories M. First, let $\mathcal{E}$ be the category of short exact sequences in M. A sequence $E \in \mathrm{Ob}\,\mathcal{E}$ can be written

$$0 \to sE \to tE \to qE \to 0.$$

We may view s, t, q as functors $\mathcal{E} \to M$.

$\mathcal{E}$ itself is an exact category, a diagram $0 \to E' \to E \to E' \to 0$ being exact if and only if the diagrams in M obtained by applying s, t, and g are exact.

Characteristic sequence theorem (4.1)   *The functor* $(s, q)\colon Q\mathcal{E} \to Q\mathcal{M} \times$
$Q\mathcal{M}$ *is a homotopy equivalence (i.e. induces a homotopy equivalence* $BQ\mathcal{E} \to$
$BQ\mathcal{M} \times BQ\mathcal{M}$).

Corollary (4.2)   *Let* $F', F, F''\colon \mathcal{M}_1 \to \mathcal{M}_2$ *be exact functors, and suppose*
natural transformations $F' \to F \to F''$ *are given such that for any object*
M_1 *of* $\mathcal{M}_1$, $0 \to F'(M_1) \to F(M_1) \to F''(M_1) \to 0$ *is exact. Then* $F_* =$
$F'_* + F''_*\colon K_*(\mathcal{M}_1) \to K_*(\mathcal{M}_2)$.

Proof   The data given amount to a functor $\mathcal{F}\colon \mathcal{M}_1 \to \mathcal{E}_2$, where $\mathcal{E}_2$ is the
category of exact sequences for $\mathcal{M}_2$. Now apply (4.1).   □

Resolution theorem (4.3)   *Let* $\mathcal{P}$ *be a full subcategory of an exact category*
$\mathcal{M}$ *which is closed under extensions and is such that*

  (i) *for any extension* $0 \to M' \to M \to M'' \to 0$ *in* $\mathcal{M}$, *if* M *is in* $\mathcal{P}$ *then* M' *is*
      in $\mathcal{P}$,
 (ii) *for any* $M'' \in \mathcal{M}$, *there exists an exact sequence as in (i) with* $M \in \mathrm{Ob}\,\mathcal{P}$.

Then $BQ\mathcal{P} \to BQ\mathcal{M}$ *is a homotopy equivalence.*

Corollary (4.4)   *Let* A *be a regular ring,* $\mathcal{M}_A$ *the category of finitely generated*
A-*modules, and* $\mathcal{P}_A$ *the category of finitely generated projective* A-*modules.*
Then $K_*(\mathcal{M}_A) \cong K_*(\mathcal{P}_A)$.

Proof   Let $\mathcal{P}_n \subset \mathcal{M}_A$ be the category of all finitely generated A-modules admitting a projective resolution of length $\le n$, so $\mathcal{P}_0 = \mathcal{P}_A$ and $\mathcal{M}_A = \cup_n \mathcal{P}_n$.
The hypotheses of (4.3) apply to $\mathcal{P}_n \subset \mathcal{P}_{n+1}$, so

$$K_*(\mathcal{P}_A) = K_*(\mathcal{P}_0) \cong K_*(\mathcal{P}_1) \cong \cdots \cong K_*(\cup_n \mathcal{P}_n) \cong U_*(\mathcal{M}_A).\qquad\square$$

Let $\mathcal{A}$ be an abelian category, $\mathcal{B} \subset \mathcal{A}$ a non-empty full abelian subcategory
closed under taking subobjects, quotient objects, and finite products in $\mathcal{A}$.

Devissage theorem (4.5)   *Suppose that every object* M *of* $\mathcal{A}$ *has a finite fil-*
tration $0 = M_0 \subset M_1 \subset \cdots \subset M_n = M$ *such that* $M_j/M_{j-1} \in \mathrm{Ob}\,\mathcal{P}$ *for all* j.
Then $K_*(\mathcal{B}) \underset{\cong}{\longrightarrow} K_*(\mathcal{A})$.

Localization theorem (4.6)   *Let* $\mathcal{B} \subset \mathcal{A}$ *be a Serre subcategory, and let* $\mathcal{A}/\mathcal{B}$
denote the quotient abelian category. Then there is a long exact sequence of
K-*groups* $\cdots \to K_n(\mathcal{B}) \to K_n(\mathcal{A}) \to K_n(\mathcal{A}/\mathcal{B}) \to K_{n-1}(\mathcal{B}) \to \cdots$.

As an example, let X be a noetherian scheme of finite Krull dimension. Let
$\mathcal{M}_X$ be the category of coherent $\mathcal{O}_X$-modules, and let $T_X^i \subset \mathcal{M}_X$ denote the full
subcategory of sheaves whose support has codimension $\ge i$. Let X^i denote the

set of points of X of codimension i. Then $T^{i+1} \subset T^i$ is a Serre subcategory, and the quotient

$$T^i/T^{i+1} \cong \coprod_{x \in X^i} \bigcup_n \mathcal{M}_{(O_{X,x}/m_{X,x}^n)},$$

where $O_{X,x}$ is the local ring at x and $m_{X,x} \subset O_{X,x}$ is the maximal ideal. Notice that an $O_{X,x}/m_{X,x}^n$-module admits a finite filtration with successive quotients $O_{X,x}/m_{X,x} = k(x)$-modules, so we may apply (4.5) and (4.6) to get (for any R a ring, $K_n(R) := K_n(\mathcal{P}_R)$, where $\mathcal{P}_R$ = f.g. projectives)

(4.7)

$$\cdots \to K_n(T^{i+1}) \longrightarrow K_n(T^i) \longrightarrow \coprod_{x \in X^i} K_n(k(x)) \longrightarrow K_{n-1}(T^{i+1}) \to \cdots$$

$$\cdots \to K_{n-1}(T^{i+2}) \to K_{n-1}(T^{i+1}) \to \coprod_{x \in X^{i+1}} K_{n-1}(k(x)) \longrightarrow \cdots .$$

The technique of exact couples shows that the above vertical maps fit together to give a complex

$$\coprod_{x \in X^0} K_n(k(x)) \to \coprod_{x \in X^1} K_{n-1}(k(x)) \to \cdots \to \coprod_{x \in X^n} K_0(k(x)),$$

which is the complex of E_1-terms of a spectral sequence

(4.8) $$E_1^{p,q} = \coprod_{x \in X^p} K_{-p-q}(k(x)) \Rightarrow K_n(\mathcal{M}_X).$$

This construction makes sense for any open set of X, and is functorial for pullback to open sets, so we obtain a spectral sequence of presheaves for the Zariski topology on X. To pass to the associated sheaves, some notation is convenient. For $p \in X^k$ and A an abelian group, let $i_p A$ denote the sheaf for the Zariski topology on X obtained by extending by zero the constant sheaf with value A on the Zariski closure of p, $\{\bar{p}\}$, to all of X. For example, if $k = \dim X$ so that p is a closed point, $i_p A$ is the skyscraper sheaf with stalk A supported at p. If $k = 0$ and X is irreducible, p is the generic point and $i_p A$ is the constant sheaf A on all of X.

For a noetherian ring or scheme we denote by K_n (resp. K_n') the K-theory of the category of finitely generated projective modules or vector bundles (resp. the category of all finitely generated modules or of coherent sheaves). On the scheme level, we define sheaves for the Zariski topology $\mathcal{K}_n$ and $\mathcal{K}_n'$ by sheafifying the presheaves

$$U \underset{\text{open}}{\subset} X \to K_n(U) \quad \text{or} \quad K'_n(U).$$

The sheaf version of (4.8) reads

(4.9) $$\mathcal{E}_1^{p,q} = \coprod_{x \in X^p} i_x K_{-p-q}(k(x)) \Rightarrow \mathcal{K}'_n.$$

The main theorem, conjectured by Gersten [7] and proved by Quillen [3], is

Theorem (4.10) *Let X be a regular k-scheme, where k is a field. Then (4.9) degenerates at E_2. In fact, $E_2^{p,q} = 0$ for $p \neq 0$, so the complex of sheaves*

$$\mathcal{K}_n \to \coprod_{x \in X^0} i_x K_n(k(x)) \to \coprod_{x \in X^1} i_x K_{n-1}(k(x)) \to \cdots \to \coprod_{x \in X^n} i_x K_0(k(x)) \to 0$$

is a resolution of the sheaf $\mathcal{K}_n$ (note $\mathcal{K}_n = \mathcal{K}'_n$ since X is regular).

Since exactness of a sequence of sheaves is a local question, the theorem can be restated:

Theorem (4.11) *Let $X = \operatorname{Spec} R$, where R is a regular local k-algebra which is a localization of a k-algebra of finite type. Then the sequence*

$$0 \to K_n(R) \to \coprod_{x \in X^0} K_n(k(x)) \to \cdots \to \coprod_{x \in X^n} K_0(k(x)) \to 0$$

is exact.

Exactness of (4.11) is easily seen to follow from

Theorem (4.12) *Let $X = \operatorname{Spec} R$ as above. Then the maps $K_*(T_X^{i+1}) \to K_*(T_X^i)$ are all zero. For the global case, the corresponding assertion is that given $\alpha \in K_*(T_X^{i+1})$ and $x \in X$, there exists an open neighborhood $U \ni x$ in X such that $\alpha \mapsto 0$ in $K_*(T_U^i)$.*

It turns out that these results can be established in a more general context which includes certain cohomology theories, as well as K-theory. I want to sketch this generalization, as well as the basic idea of the proof. Let X be a variety over a field k, and consider a "cohomology theory" $H^*(X)$ which may be one of the following:

(i) $k = \mathbf{C}$, $H^*(X) = H^*(X, A)$, ordinary singular cohomology with coefficients in some abelian group A (e.g. $A = \mathbf{Z}, \mathbf{Q}, \mathbf{R}$, or $\mathbf{C}$),

(ii) $\operatorname{char} k = 0$ and $H^*(X) = H^*_{\mathrm{DR}}(X)$, de Rham cohomology [10],

(iii) r prime to $\operatorname{char} k$ and $H^*(X) = H^*_{\mathrm{et}}(X, \mu_n^{\otimes r})$, étale cohomology with coefficients in the rth twist of the sheaf μ_n of nth roots of 1.

The role of $\mu_n^{\otimes r}$ in case (iii) is confusing. For purposes of exposition we will fix an nth root of 1, $\xi \in k$, and identify $\mu_n^{\otimes r} \cong \mathbf{Z}/n\mathbf{Z}$. Then at the end we will simply state what the result would be, keeping track of the twist by r.

If $Z \subset X$ is a closed subvariety of codimension p, there is (in any of the three cases above) a notion of cohomology with supports in Z, $H_Z^*(X)$, fitting into a long exact sequence

$$(4.13) \qquad \cdots \to H_Z^r(X) \to H^r(X) \to H^r(X - Z) \to H_Z^{r+1}(X) \to \cdots .$$

(A topologist would write $H_Z^r(X) = H^r(X, X - Z)$.) One of the basic tenets of Grothendieck duality theory (Verdier [8], [9]) is that if X is smooth over k of dimension n, then for any of the above theories, if we write

$$H_r(Z) := H_Z^{2n-r}(X),$$

then $H_r(Z)$ is independent of X and covariant functorial for proper maps $Z \to Z'$. In fact, $H_*(Z)$ is a *Borel–Moore homology theory* [9] in the sense that if Z is itself smooth (but *not* necessarily complete) of dimension d, then $H_r(Z) \cong H^{2d-r}(Z)$ (simply take $Z = X$ in the above discussion).

As a good exercise for understanding the various exact sequences, the reader should construct an exact sequence (for $Z' \subset Z \subset X$ closed subschemes, with X smooth)

$$(4.14) \qquad \cdots \to H_r(Z') \to H_r(Z) \to H_r(Z - Z') \to H_{r-1}(Z') \to \cdots .$$

Now write

$$H_{Z^p}^*(X) = \varinjlim_{\substack{Z \subset X \\ \text{codim } Z \geq p}} H_Z^*(X)$$

and deduce from (4.14) long exact sequences (compare with (4.7))

$$(4.15) \quad \cdots \to H_{Z^{i+1}}^r(X) \to H_{Z^i}^r(X) \longrightarrow \coprod_{x \in X^i} H^{r-2i}(k(x)) \longrightarrow H_{Z^{i+1}}^{r+1}(X) \to \cdots$$

$$\Big\downarrow =$$

$$\cdots \to H_{Z^{i+2}}^{r+1}(X) \to H_{Z^{i+1}}^{r+1}(X) \to \coprod_{x \in X^{i+1}} H^{r-2i-1}(k(x)) \longrightarrow \cdots .$$

(Here if $x \in Z \subset X$ is the generic point of Z, then we define $H^*(k(x)) := \varinjlim_{U \subset Z, \text{ open}} H^*(U)$.) Precisely as before, one can use the exact couple technique to construct a spectral sequence

$$(4.16) \qquad E_1^{p,q} = \coprod_{x \in X^p} H^{q-p}(k(x)) \Rightarrow H^{p+q}(X).$$

The filtration $F^* H^*(X)$ deduced from (4.16) is the filtration by codimension of support; that is, $a \in F^p H^*(X)$ if and only if there exists $Z \subset X$ a subscheme of codimension p such that $a \mapsto 0$ in $H^*(X - Z)$. Also, the complex of E_1-terms (4.16) can be sheafified as in (4.10), and one has

Theorem (4.17)   *The complex of sheaves*

$$\coprod_{x \in X^0} i_x H^p(k(x)) \to \coprod_{x \in X^1} i_x H^{p-1}(k(x)) \to \cdots \to \coprod_{x \in X^p} i_x H^0(k(x)) \to 0$$

is a resolution of the Zariski sheaf $\mathcal{H}^p$ associated to the presheaf $U \to H^p(U)$.

Corollary (4.18)   *The E_2-term of (4.16) is $E_2^{p,q} = H_{Zar}^p(X, \mathcal{H}^q)$. We have $H_{Zar}^p(X, \mathcal{H}^q) = 0$ for $p > q$.*

Proof   The sheaves in (4.17) have no higher cohomology, so the complex of abelian groups calculates $H^*(X, \mathcal{H}^p)$.   □

Remark (4.19)   For either the étale or singular cohomology $H^*(X)$, one calculates with a finer topology on X than the Zariski topology. One has a morphism of *sites* $X_{\text{fine}} \xrightarrow{\pi} X_{Zar}$ and hence a spectral sequence $H_{Zar}^p(X, R^q \pi_*) \Rightarrow H_{\text{fine}}^{p+q}(X)$. This spectral sequence coincides with (4.16) from E_2 onward. For a proof in the de Rham theory, see Bloch and Ogus [6]. The general result is due to Deligne.

Corollary (4.20)   *One has*

$$H^p(X, \mathcal{K}_p) \cong CH^p(X);$$

$$H^p(X, \mathcal{H}^p) \cong CH^p(X)/A^p(X), \quad \text{for } X \text{ over } \mathbf{C}, \ H^*(X) = H^*(X, \mathbf{Z});$$

$$H^p(X, \mathcal{H}^p) \cong CH^p(X)/\ell CH^p(X),$$

$$\text{for } X \text{ over } k = \bar{k}, \ H^*(X) = H_{\text{et}}^*(X, \mathbf{Z}/\ell \mathbf{Z}) \ (\ell, \operatorname{char} k) = 1.$$

Proof   One has (because the complexes of global sections of (4.10) and (4.17) calculate the cohomology)

$$(4.21) \qquad \coprod_{x \in X^{p-1}} k(x)^* \to \coprod_{x \in X^p} \mathbf{Z} \to H^p(X, \mathcal{K}_p) \to 0,$$

using that $K_1(k(x)) = k(x)^*$ and $K_0(k(x)) = \mathbf{Z}$, and

$$(4.22) \qquad \coprod_{x \in X^{p-1}} H^1(k(x)) \to \coprod_{x \in X^p} H^0(k(x)) \to H^p(X, \mathcal{H}^p) \to 0.$$

The idea of the proof for K-theory is that an irreducible family of codimension-p cycles parameterized by $\mathbf{P}^1$

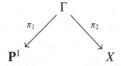

gives rise to a codimension-$(p-1)$ subvariety $\pi_2(\Gamma)$ and norm$_{k(\Gamma)/k(\pi_2\Gamma)}(t)$ on $\pi_2(\Gamma)$, where $t = \pi_1^*(u)$ and u is the "standard" function on $\mathbf{P}^1$ ($u(0) = 0, u(\infty) = \infty$). For more details, see Quillen [3].

For the étale theory, $H^1(k(x)) \cong H^1_{\text{Gal}}(k(x), \mu_\ell)$ (Galois cohomology). This group can be calculated by the Kümmer sequence

$$0 \to \mu_\ell \to \overline{k(x)}^* \xrightarrow{\ell} \overline{k(x)}^* \to 0 \qquad (\overline{k(x)} = \text{algebraic closure of } k(x))$$

and the known vanishing of $H^1_{\text{Gal}}(k(x), \overline{k(x)}^*)$ (Hilbert's theorem 90). One gets $H^1(k(x)) \cong k(x)^*/k(x)^{*\ell}$. Since $H^0(k(x)) = \mathbf{Z}/\ell\mathbf{Z}$, the presentation (4.22) for the étale theory becomes

$$\coprod_{X^{p-1}} k(x)^*/k(x)^{*\ell} \to \coprod_{X^p} \mathbf{Z}/\ell\mathbf{Z} \to H^p(X, \mathcal{H}^p) \to 0,$$

whence $H^p(X, \mathcal{H}^p) \cong CH^p(X)/\ell CH^p(X)$.

Finally, for the singular theory $H^*(X) = H^*(X, \mathbf{Z})$ one uses the fact that algebraic and homological equivalence coincide for divisors to establish

$$CH^1(\Gamma)/A^1(\Gamma) \cong H^1_{\text{Zar}}(\Gamma, \mathcal{H}^1) \subset H^2_{\text{sing}}(\Gamma, \mathbf{Z})$$

for any complete nonsingular Γ. Now for a diagram

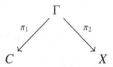

as before, where C now is any smooth connected curve and Γ is smooth (using resolution of singularities, we may assume this) one gets a commutative diagram

$$\begin{array}{ccccccc}
H^1(k(\Gamma)) & \longrightarrow & \coprod_{\gamma \in \Gamma^1} \mathbf{Z} & \longrightarrow & H^1(\Gamma, \mathcal{H}^1) & \longrightarrow & 0 \\
\downarrow {\scriptstyle \text{norm}} & & \downarrow {\scriptstyle \text{norm}} & & \downarrow & & \\
\coprod_{X^{p-1}} H^1(k(x)) & \longrightarrow & \coprod_{X^p} \mathbf{Z} & \longrightarrow & H^p(X, \mathcal{H}^p) & \longrightarrow & 0.
\end{array}$$

The desired isomorphism follows from this. For more details, the reader should see Bloch and Ogus [6].   □

We now sketch a proof of (4.17), indicating at the end how a similar argument proves (4.10). First, Theorem (4.17) can be reformulated in the spirit of (4.12). Given $\alpha \in H^r_{Z^p}(X)$ and $x \in X$, it suffices to show there exists an open neighborhood $x \in U \subset X$ such that $\alpha \mapsto 0$ in $H^r_{Z^{p-1}}(U)$. Now, let $n = \dim X$ and let $Z \subset X$ be a subscheme of pure codimension p such that α lies in the image of $H_{2n-r}(Z) \to H^r_{Z^p}(X)$. It suffices to exhibit an open neighborhood U of x and a closed subscheme $Z' \subset U$ of codimension $p - 1$ such that $Z \cap U \subset Z'$ and such that the map $H_{2n-r}(Z) \to H_{2n-r}(Z')$ is zero.

As a first step, replacing X by an open neighborhood of x and Z by its intersection with this neighborhood, one can construct a cartesian diagram (by a variant on Noether normalization)

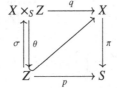

with $S \subset \mathbf{A}^{n-1}$ an open subscheme, π smooth with fibre dimension 1, and p finite. The pullback θ of π will then also be smooth, so by further localization one can arrange for the image of Z under the section σ, $\sigma(Z) \subset X \times_S Z$, to be defined by the vanishing of a single function $\sigma(Z) : f = 0$. Let $Z' = q(X \times_S Z) \subset X$. We have

$$H_*(Z) \xrightarrow{\sigma_*} H_*(X \times_S Z) \xrightarrow{q_*} H_*(Z')$$

so it suffices to show $\sigma_* = 0$. The best way to think of this is topologically.

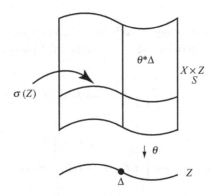

If $\Delta \in H_*(Z)$, then $\sigma_*(\Delta) = \theta^* \Delta \cap [\sigma(Z)]$, where $\theta^* : H_*(Z) \to H_{*+2}(X \times_S Z)$ is topological pullback (which exists even though θ is not proper), $\cap$ is cap product, and $[\sigma(Z)] \in H^2(X \times_S Z)$ is the divisor class. By hypothesis $\sigma(Z)$: $f = 0$, so $[\sigma(Z)] = 0$ in $H^2(X \times_S Z)$ and $\sigma_* = 0$ as claimed. For complete details of these arguments, the reader is referred to Bloch and Ogus [6].

Remark (4.23) Quillen's proof of (4.12) differed only in detail from the above. Given X a regular scheme of finite type over a field k, $x \in X$, and $\alpha \in K_*(T_X^{i+1})$, one localizes around x to achieve a diagram

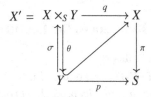

where Y is a divisor in X, $\alpha \in \text{Image}(K_*(T_Y^i) \to K_*(T_X^{i+1}))$, and π smooth with fibre dimension 1 and p finite. Localizing further, one can arrange (since θ is smooth with fibre dimension 1) for $\sigma(Y) : f = 0$ in X'. Writing $S = \text{Spec } A$, $Y = \text{Spec } B$, $X = \text{Spec } C$, $C' = C \otimes_A B$, one has $K_*(T_B^i) \xrightarrow{\sigma} K_*(T_{C'}^i) \xrightarrow{q_*} K_*(T_C^i)$ and one wants to show $\sigma_* = 0$.

The exact sequence

$$0 \to C' \xrightarrow{\cdot f} C' \underset{\theta^*}{\overset{\sigma^*}{\rightleftarrows}} B \to 0$$

is split, and θ^* is flat, so for any $M \in \text{Ob } T_B^i$ one gets

$$0 \to C' \otimes_B M \xrightarrow{\cdot f} C' \otimes_B M \to M \to 0.$$

In other words, there is an exact sequence

$$0 \to \theta^* \xrightarrow{\cdot f} \theta^* \to \sigma_* \to 0$$

of exact functors $T_B^i \to T_{C'}^i$. One now applies (4.1)

$$\theta^* = \theta^* + \sigma_* : K_*(T_B^i) \to K_*(T_{C'}^i),$$

that is $\sigma_* = 0$.

Remark (4.24) For ease of exposition, I fixed an identification $\mu_n \cong \mathbf{Z}/n\mathbf{Z}$ in working with the étale theory. This can be avoided. Keeping track of the twisting, the spectral sequence (4.16) becomes

$$E_1^{p,q} = \coprod_{x \in X^p} H_{\text{Gal}}^{q-p}\left(k(x), \mu_n^{\otimes m-p}\right) \Rightarrow H_{\text{et}}^{p+q}\left(X, \mu_n^{\otimes m}\right),$$

and the complex (4.17) reads

$$\coprod_{x \in X^0} i_x \, H^p\left(k(x), \mu_n^{\otimes m}\right) \to \coprod_{x \in X^1} i_x \, H^{p-1}\left(k(x), \mu_n^{\otimes m-1}\right) \to \cdots$$

$$\to \coprod_{x \in X^p} i_x \, H^0\left(k(x), \mu_n^{\otimes m-p}\right) \to 0.$$

This complex resolves the sheaf $\mathcal{H}^p(\mu_n^{\otimes m})$ associated to the presheaf $U \mapsto H_{\text{et}}^p(U, \mu_n^{\otimes m})$.

References for Lecture 4

Foundational works on K-theory:

[1] H. Bass, *Algebraic K-Theory*, Benjamin, New York (1968).

[2] J. Milnor, *Introduction to Algebraic K-Theory*, Annals of Mathematics Studies, vol. 72, Princeton University Press, Princeton, N.J. (1971).

[3] D. Quillen, Higher algebraic K-theory. I, in *Algebraic K-Theory I*, Lecture Notes in Math., no. 341, Springer, Berlin (1973).

[4] D. Quillen and D. Grayson, Higher algebraic K-theory. II, pp. 217–240 in *Algebraic K-Theory*, Lecture Notes in Math., no. 551, Springer, Berlin (1976).

[5] R. Swan, *Algebraic K-Theory*, Lecture Notes in Math., no. 76, Springer, Berlin (1968).

Work on Gersten's conjecture and analogous results for cohomology theories:

[6] S. Bloch and A. Ogus, Gersten's conjecture and the homology of schemes, *Ann. Sci. École Norm. Sup. (4)*, **7** (1974), 181–201 (1975).

[6a] S. Bloch, K_2 and algebraic cycles, *Ann. of Math. (2)*, **99** (1974), 349–379.

[7] S. M. Gersten, Some exact sequences in the higher K-theory of rings, pp. 211–243 in *Algebraic K-Theory I*, Lecture Notes in Math., no. 341, Springer, Berlin (1973).

Some expository references for various cohomology theories and Grothendieck duality:

[8] J.-L. Verdier, A duality theorem in the etale cohomology of schemes, pp. 184–198 in *Proceedings of a conference on local fields (Driebergen, 1966)*, Springer, Berlin (1967).

[9] J.-L. Verdier, Dualité dans la cohomologie des espaces localement compacts, *Séminaire Bourbaki, Vol. 9*, exposé no. 300 (1965), 337–349. [Reprinted Société Mathématique de France, Paris, 1995.]

[10] R. Hartshorne, On the de Rham cohomology of algebraic varieties, *Inst. Hautes Études Sci. Publ. Math.*, no. 45 (1975), 5–99.

[11] P. Deligne, *Cohomologie étale* (SGA $4\frac{1}{2}$), Lecture Notes in Math., no. 569, Springer, Berlin (1977).

5

Torsion in the Chow group

The purpose of this lecture is to apply the techniques of Lecture 4 in order to prove

Theorem (5.1) *Let X be a smooth projective variety over an algebraically closed field k. Then the map $A_0(X) \to \mathrm{Alb}(X)$ induces an isomorphism on torsion prime to the characteristic of k.*

This theorem was first proved geometrically by A. A. Roitman. For a different geometric proof, see Bloch [3].

We have seen already (Lecture 1, proof of Lemma (1.4)) that the map on N-torsion $_NA_0(X) \to {}_N\mathrm{Alb}(X)$ is surjective for any N prime to char k. (The proof given there applies to $k = \mathbf{C}$, but the same discussion is valid quite generally with singular cohomology replaced by étale.)

Lemma (5.2) *Suppose the map $_NA_0(X) \to {}_N\mathrm{Alb}(X)$ is injective for any surface X. Then it is so for X of arbitrary dimension.*

Proof Let dimension $X > 2$, and suppose we are given a zero-cycle

$$\delta = \sum n_i (x_i)$$

on X such that $N\delta$ is rationally equivalent to zero. By definition, there exist irreducible curves $C_i \subset X$ and functions f_i on C_i such that $\sum(f_i) = N\delta$.

By a succession of blowings up of nonsingular points on X, one constructs $X' \xrightarrow{\pi} X$ such that the strict transform C'_i of C_i on X' is nonsingular for all i, and such that moreover no two C'_i meet. Let $\delta' = \sum n_i (X'_i)$ be some lifting of δ to X' (i.e. $\pi_*\delta' = \delta$). Then viewing the f_i as functions on C'_i, we get $\sum(f_i) = N\delta' + \eta$, where $\pi_*(\eta) = 0$. Since the exceptional divisor in X' is a union of projective spaces, it is easy to see that there will exist lines $\ell_j \subset X'$ and functions g_j on ℓ_j such that $\pi(\ell_j) = \mathrm{pt}$, $\sum(g_j) = \eta$, and no more than two ℓ_j meet at a point.

Let $D = \cup C'_i \cup \cup \ell_j$. Let $Y \subset X$ be a hypersurface section of large degree.

Suppose $D \subset Y$ but Y is otherwise "general." I claim that such a Y is nonsingular. Let $O_{X'}(1)$ be the ample bundle on X' and let Y correspond to a section $\sigma \in \Gamma(X', O_{X'}(d))$. Let I be the ideal of $D \subset X'$. We may assume $d \gg 0$ so $I(d)$ is generated by its global sections. Since σ is general, this implies by a standard Bertini argument that Y is nonsingular off of D. Further we may assume $\Gamma(X', I(d)) \to \Gamma(X', (I/I^2)(d))$ is surjective and the sheaf $(I/I^2)(d)$ is generated by global sections. Since $\operatorname{rk} I/I^2 > 1$ and this sheaf is locally free off the singular points of D, the section of $(I/I^2)(d)$ induced by σ vanishes at worst at singular points of D. This implies that Y is only singular (possibly) at D_{sing}. Finally, since a singular point of D consists of two simple branches, it will have embedding dimension ≤ 2. Since $\dim Y \geq 2$, one sees easily that Y can be chosen nonsingular.

We may continue this construction inductively, cutting Y by a hypersurface section, etc., until we get

$$D \subset Z \subset X,$$

where $Z \subset X$ is a smooth complete intersection of hypersurfaces with $\dim Z = 2$. Notice the cycle δ' is supported on Z, and $N\delta'$ is rationally equivalent to 0 (on Z!). One has well-known isomorphisms (cf. Kleiman [8])

$$\mathrm{Alb}(Z) \cong \mathrm{Alb}(X') \cong \mathrm{Alb}(X)$$

and a diagram

$$
\begin{array}{ccc}
\delta' \quad \in \quad {}_N A_0(Z) & \longrightarrow & {}_N \mathrm{Alb}(Z) \\
\downarrow & \downarrow & \downarrow \cong \\
\delta \quad \in \quad {}_N A_0(X) & \longrightarrow & {}_N \mathrm{Alb}(X)
\end{array}
$$

Assuming the top arrow injective, we see $\delta \mapsto 0 \Rightarrow \delta = 0$. Since δ was arbitary, this proves the lemma. $\square$

Lemma (5.3) *To establish injectivity ${}_N A_0(X) \to {}_N \mathrm{Alb}(X)$, it suffices to consider the case $N = \ell = $ prime.*

Proof This is straightforward, using divisibility of $A_0(X)$ (Lecture 1, (1.3)). $\square$

We assume henceforth $N = \ell = $ prime and $\dim X = 2$. We want to prove ${}_\ell A_0(X) \to {}_\ell \mathrm{Alb}(X)$. We work with the reduction mod ℓ of the sheaf $\mathcal{K}_{2_X}$ discussed in Lecture 4.

Lemma (5.4) *There is an exact sequence (Zariski cohomology)*

$$0 \to H^1(X, \mathcal{K}_2)/\ell^\nu H^1(X, \mathcal{K}_2) \to H^1(X, \mathcal{K}_2/\ell^\nu \mathcal{K}_2) \to {}_{\ell^\nu} A_0(X) \to 0.$$

Proof The rows and columns of the following diagram of sheaves are exact

(5.4.1)

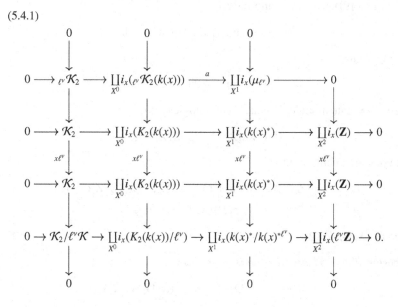

This is not quite obvious, but becomes so once one observes that the map labeled a on the top row is surjective. For example one can tensor the divisor map $\coprod_{X^0} i_x(k(x)^*) \twoheadrightarrow \coprod_{X^1} i_x \mathbf{Z}$ with μ_{ℓ^v} and get

$$\coprod_{X^0} i_x(k(x)^* \otimes \mu_{\ell^v}) \twoheadrightarrow \coprod_{X^1} i_x \mu_{\ell^v}$$

$$\coprod_{X^0} i_x({}_{\ell^v} K_2(k(x))).$$

with diagonal map a.

The first row of (5.4.1) implies ${}_{\ell^v}\mathcal{K}_2$ has cohomological dimension ≤ 1, and the first column now gives the desired sequence, using ${}_{\ell^v}A_0(X) \cong {}_{\ell^v}\mathrm{CH}_0(X) \cong {}_{\ell^v}H^2(X, \mathcal{K}_2)$. □

We can, if we like, pass to the limit in (5.4), getting

$$0 \to H^1(X, \mathcal{K}_2) \otimes \mathbf{Q}_\ell/\mathbf{Z}_\ell \to H^1(X, \mathcal{K}_2 \otimes \mathbf{Q}_\ell/\mathbf{Z}_\ell) \to A_0(X)(\ell) \to 0,$$

where $A_0(X)(\ell)$ denotes the ℓ-power torsion subgroup.

Lemma (5.5) $H^3_{\mathrm{et}}(X, \mathbf{Q}_\ell/\mathbf{Z}_\ell(2)) \cong \mathrm{Alb}(X)(\ell)$.

Proof   The Albanese and Picard varieties of X are dual, so the e_m-pairing of Weil (Lang [9]) gives an isomorphism

$$_{\ell^\nu}\text{Alb}(X) \cong \text{Hom}(_{\ell^\nu}\text{Pic Var}(X), \mu_{\ell^\infty}).$$

Passing to the limit over ν

$$\text{Alb}(X)(\ell) \cong \text{Hom}(T_\ell(\text{Pic Var}(X)), \mu_{\ell^\infty}),$$

where T_ℓ = Tate module. In étale cohomological terms

$$T_\ell(\text{Pic Var}(X)) \cong H^1_{\text{et}}(X, \mathbf{Z}_\ell(1)),$$

and Poincaré duality implies

$$H^3_{\text{et}}(X, \mathbf{Q}_\ell/\mathbf{Z}_\ell(2)) \cong \text{Hom}(H^1(X, \mathbf{Z}_\ell(1)), H^4(X, \mathbf{Q}_\ell/\mathbf{Z}_\ell(3)))$$
$$\cong \text{Hom}(H^1(X, \mathbf{Z}_\ell(1)), \mu_{\ell^\infty})$$

since $H^4(X, \mathbf{Q}_\ell/\mathbf{Z}_\ell(2)) \cong \mathbf{Q}_\ell/\mathbf{Z}_\ell$. The lemma now follows.   □

Theorem (5.6)   *There exists an isomorphism*

$$\alpha \colon H^1(X, \mathcal{K}_2 \otimes \mathbf{Q}_\ell/\mathbf{Z}_\ell) \cong H^3_{\text{et}}(X, \mathbf{Q}_\ell/\mathbf{Z}_\ell(2))$$

making the diagram below commute:

$$
\begin{array}{ccccc}
H^1(X, \mathcal{K}_2 \otimes \mathbf{Q}_\ell/\mathbf{Z}_\ell) & \longrightarrow & A_0(X)(\ell) & \longrightarrow & 0 \\
\alpha \downarrow & & \downarrow & & \\
H^3_{\text{et}}(X, \mathbf{Q}_\ell/\mathbf{Z}_\ell(2)) & \underset{\approx}{\longrightarrow} & \text{Alb}(X)(\ell). & & \\
& & \downarrow & & \\
& & 0 & &
\end{array}
$$

Note that the right-hand vertical arrow above is the cycle map, so Theorem (5.1) will follow from (5.6).

We prove (5.6) by constructing isomorphisms for any ν

$$\alpha_\nu \colon H^1(X, \mathcal{K}_2/\ell^\nu\mathcal{K}_2) \underset{\cong}{\longrightarrow} H^3_{\text{et}}(X, \mu_{\ell^\nu}^{\otimes 2}).$$

The first step is to interpret H^3_{et} as a Zariski cohomology group. Recall from Lecture 4, the Zariski sheaves $\mathcal{H}^q(\mu_{\ell^\nu}^{\otimes 2})$ on X were defined to be the sheaves associated to the presheaves $U \mapsto H^q(X, \mu_{\ell^\nu}^{\otimes 2})$. This implies that $\mathcal{H}^q$ is the qth right derived functor of the morphism of topoi $\pi \colon X_{\text{et}} \to X_{\text{Zar}}$, so there is a Leray spectral sequence

$$E_2^{p,q} = H^p_{\text{Zar}}(X, \mathcal{H}^q(\mu_{\ell^\nu}^{\otimes 2})) \Rightarrow H^{p+q}_{\text{et}}(X, \mu_{\ell^\nu}^{\otimes 2}).$$

In our case $\mathcal{H}^q = 0$ for $q \geq 3$ because $\dim X = 2$, so locally (say on pieces of an affine covering) X has cohomological dimension 2. Also we wrote down an acyclic resolution of $\mathcal{H}^q$ of length q, so $H^p(X, \mathcal{H}^q) = 0$ for $p > q$. The above spectral sequence looks like

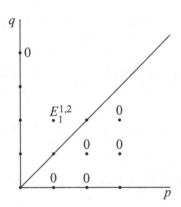

from which $E_2^{1,2} = H^1_{\mathrm{Zar}}(X, \mathcal{H}^2(\mu_{\ell^\nu}^{\otimes 2})) \cong H^3_{\mathrm{et}}(X, \mu_{\ell^\nu}^{\otimes 2})$.

Our objective thus becomes to construct an isomorphism

$$(5.6.1) \qquad \alpha_\nu : H^1(X, \mathcal{K}_2/\ell^\nu \mathcal{K}_2) \xrightarrow{\cong} H^1(X, \mathcal{H}^2(\mu_{\ell^\nu}^{\otimes 2})).$$

We will actually construct an exact sequence of sheaves

$$(5.6.2) \qquad 0 \to C \to \mathcal{K}_2/\ell^\nu \mathcal{K}_2 \to \mathcal{H}^2(\mu_{\ell^\nu}^{\otimes 2}) \to 0,$$

where C is a constant sheaf. (It may well be that $C = 0$. This is known when $\ell = 2$; Elman and Lam [7].)

Recall Tate's construction of the Galois symbol [10]

$$h : K_2(k(X))/\ell^\nu K_2(k(X)) \to H^2_{\mathrm{Gal}}(k(X), \mu_{\ell^\nu}^{\otimes 2}).$$

We have seen in the proof of (4.20) that $H_{\mathrm{Gal}}(k(X), \mu_{\ell^\nu}) \cong k(X)^*/k(X)^{*\ell^\nu}$, so there is a cup product pairing

$$k(X)^*/k(X)^{*\ell^\nu} \times k(X)^*/k(X)^{*\ell^\nu} \to H^2_{\mathrm{Gal}}(k(X), \mu_{\ell^\nu}^{\otimes 2}).$$

Define $h\{f, g\} = \bar{f} \cup \bar{g}$ where bar denotes class modulo ℓ^ν-th powers. To see $\bar{f} \cup \overline{(1 - f)} = 0$, note $1 - f$ is a norm from the field $L = k(X)(f^{\ell^{-\nu}})$, so $\overline{(1 - f)} = \mathrm{cor}_{L/k(X)}(a)$ for some $a \in H^1_{\mathrm{Gal}}(L, \mu_{\ell^\nu})$. Using the projection formula

$$\bar{f} \cup \overline{(1 - f)} = \bar{f} \cup \mathrm{cor}(a) = \mathrm{cor}_{L/k(X)}(\bar{f} \cup a) = \ell^\nu \mathrm{cor}(\overline{f^{\ell^{-\nu}}} \cup a) = 0.$$

The Galois symbol relates the solutions of $\mathcal{K}_2/\ell^\nu\mathcal{K}_2$ (5.4.1) and $\mathcal{H}^2$ (4.17) as follows:

(5.6.3)

$$
\begin{array}{ccc}
0 & & 0 \\
\downarrow & & \downarrow \\
K_2/\ell^\nu K_2 \dashrightarrow \mathcal{H}^2(\mu_{\ell^\nu}^{\otimes 2}) & & \\
\downarrow & & \downarrow \\
\coprod_{X^0} i_x(K_2(k(x))/\ell^\nu K_2(k(x))) \xrightarrow{\ h\ } \coprod_{X^0} i_x(H^2_{\mathrm{Gal}}(k(x),\mu_{\ell^\nu}^{\otimes 2})) \\
\downarrow & & \downarrow \\
\coprod_{X^1} i_x(k(x)^*/k(x)^{*\ell^\nu}) =\!\!=\!\!= \coprod_{X^1} i_x(k(x)^*/k(x)^{*\ell^\nu}) \\
\downarrow & & \downarrow \\
\coprod_{X^2} i_x(\mathbf{Z}/\ell^\nu\mathbf{Z}) =\!\!=\!\!= \coprod_{X^2} i_x(\mathbf{Z}/\ell^\nu\mathbf{Z}) \\
\downarrow & & \downarrow \\
0 & & 0.
\end{array}
$$

The reader can show as an exercise that the squares of (5.6.3) commute. We will thus have our exact sequence (5.6.2) and the ball game will be over, once we show

Theorem (5.7) $h: K_2(k(X)) \to H^2_{\mathrm{Gal}}(k(X),\mu_{\ell^\nu}^{\otimes 2})$ *is surjective.*

Proof   Note that $k(X)$ has Galois cohomological dimension 2, so there is a diagram

$$
\begin{array}{ccccccc}
K_2(k(X)) & \xrightarrow{\ \ell\ \ } & K_2(k(X)) & \longrightarrow & K_2(k(X))/\ell_2(k(X)) & \longrightarrow & 0 \\
\downarrow{\scriptstyle h} & & \downarrow{\scriptstyle h} & & \downarrow{\scriptstyle h} & & \\
H^2(k(X),\mu_{\ell^{\nu-1}}^{\otimes 2}) & \longrightarrow & H^2(k(X),\mu_{\ell^\nu}^{\otimes 2}) & \longrightarrow & H^2(k(X),\mu_\ell^{\otimes 2}) & \longrightarrow & 0.
\end{array}
$$

From this, one reduces to the case $\nu = 1$.

Let $f: X \to \mathbf{P}^1$ be a dominant rational map. Let $F = k(X)$, $K = k(\mathbf{P}^1) \subset F$, and let Y be an affine open subvariety of the generic fibre of f. The subvariety Y is thus an open curve defined and of finite type over K. We assume Y is geometrically reduced and connected (i.e. K is algebraically closed in F).

Let $K' \supset K$ be a finite extension field, $Y' = Y_{K'}$, $F' = F \cdot K'$, $\pi: Y' \to Y$ the natural finite morphism. Associated to π there is a covariant trace morphism π_* on étale cohomology, which is analogous to the corestriction on Galois cohomology. In fact, there is a commutative diagram

$$(5.7.1) \quad \begin{array}{ccc} H^r_{\mathrm{et}}(Y',\mu_\ell^{\otimes 2}) & \longrightarrow & H^r_{\mathrm{Gal}}(F',\mu_\ell^{\otimes 2}) \\ {\scriptstyle \pi_*}\downarrow & & \downarrow{\scriptstyle \mathrm{cor}} \\ H^r_{\mathrm{et}}(Y,\mu_\ell^{\otimes 2}) & \longrightarrow & H^r_{\mathrm{Gal}}(F,\mu_\ell^{\otimes 2}). \end{array}$$

One way to understand this situation is to notice that the "geometric fibres" of π (i.e. schemes of the form $Y' \times_Y \operatorname{Spec} A$, where A is a strictly Henselian ring lying over some local ring of Y) are finite over $\operatorname{Spec} A$ and hence strictly Henselian. This implies that the higher derived functors of π are trivial, so that $H^r(Y',\mu_\ell^{\otimes 2}) \cong H^r(Y, \pi_*(\mu_\ell^{\otimes 2}))$. The map π_* comes from a map on sheaves $\pi_*(\mu_{\ell Y'}^{\otimes 2}) \to \mu_{\ell Y}^{\otimes 2}$, which on the fibres is simply summing over connected components. In particular, the map on sheaves is surjective. Since Y has cohomological dimension 2, it follows that π_* in (5.7.1) is surjective.

We were free initially to shrink (localize) Y as much as we wanted, so it will suffice to show

$$\operatorname{Image}(H^2(Y,\mu_\ell^{\otimes 2}) \to H^2(F,\mu_\ell^{\otimes 2})) \subset \operatorname{Image}(K_2(F) \to H^2(F,\mu_\ell^{\otimes 2})).$$

There is also a transfer map tr: $K_2(F') \to K_2(F)$, and the diagram

$$(5.7.2) \quad \begin{array}{ccc} K_2(F') & \xrightarrow{\ h\ } & H^2(F',\mu_\ell^{\otimes 2}) \\ {\scriptstyle \mathrm{tr}}\downarrow & & \downarrow{\scriptstyle \mathrm{cor}} \\ K_2(F) & \xrightarrow{\ h\ } & H^2(F,\mu_\ell^{\otimes 2}) \end{array}$$

is known to commute (Bass and Tate [2]). Together with (5.7.1) and the surjectivity of π_*, this reduces us to proving

$$\operatorname{Image}(H^2(Y',\mu_\ell^{\otimes 2}) \to H^2(F',\mu_\ell^{\otimes 2})) \subset \operatorname{Image}(K_2(F') \to H^2(F',\mu_\ell^{\otimes 2}))$$

for some finite extension K' of K.

Let $\bar{K}$ be the separable closure of K and write $\bar{Y} = Y_{\bar{K}} \cdot \bar{Y}$ has cohomological dimension 1, and the Hochschild–Serre spectral sequence implies

$$H^2(Y,\mu_\ell^{\otimes 2}) \cong H^1_{\mathrm{Gal}}(K, H^1(\bar{Y},\mu_\ell^{\otimes 2})).$$

The point is (and this explains why we must work with varieties and étale cohomology rather than the more familiar Galois cohomology of fields) that $H^1(\bar{Y},\mu_\ell^{\otimes 2})$ is a finite group. Thus for some finite K' over K we may suppose

(i) $\operatorname{Gal}(\bar{K}/K')$ acts trivially on $H^1(\bar{Y},\mu_\ell^{\otimes 2})$;
(ii) $H^1(Y',\mu_\ell^{\otimes 2})$ maps onto $H^1(\bar{Y},\mu_\ell^{\otimes 2})$.

By the previous discussion, we can replace Y and Y' and assume (i) and (ii) hold for Y and K. We then have

(5.7.3)

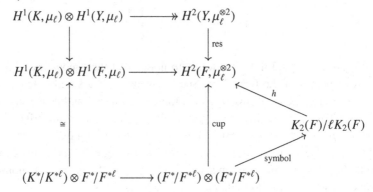

The reader easily checks that Image(res) $\subset$ Image(h) as claimed.   □

Remark (5.8)   The arguments in this lecture can be pushed further in several directions. For a surface X over an algebraically closed field of characteristic prime to ℓ, one can show that the ℓ-power torsion in $K_1(X)$, $K_1(X)(\ell)$, maps onto $H^2_{et}(X, \mathbf{Q}_\ell / \mathbf{Z}_\ell(2))$, the map being an isomorphism at least when $\ell = 2$. This leads to some detailed conjectures concerning Chern characters linking K-theory with finite coefficients and étale cohomology. I hope to publish a description of these conjectures and the evidence for them elsewhere.

Concerning (5.7), one can show by a similar argument that for F a function field of transcendence degree n over an algebraically closed field, the cup product

$$H^1(F, \mu_{\ell^r})^{\otimes n} \xrightarrow{\;\;\vee\;\;} H^n(F, \mu_{\ell^r}^{\otimes n})$$

is surjective.

I wonder whether the whole cohomology algebra $\oplus H^r(F, \mu_{\ell^r}^{\otimes r})$ might not be generated by H^1? If true, this would imply (identifying $\mu \cong \mathbf{Z}/\ell\mathbf{Z}$) that the Bockstein maps

$$H^r(F, \mathbf{Z}/\ell\mathbf{Z}) \to H^{r+1}(F, \mathbf{Z}/\ell\mathbf{Z})$$

were all zero. Globally this would imply the existence of a divisor $D \subset X$ such that the composition

$$H^r_{et}(X, \mathbf{Z}/\ell\mathbf{Z}) \xrightarrow{\text{Bockstein}} H^{r+1}_{et}(X, \mathbf{Z}/\ell\mathbf{Z}) \to H^{r+1}_{et}(X - D, \mathbf{Z}/\ell\mathbf{Z})$$

was zero. In other words, the image of the Bockstein would lie in the first level of the "coniveau filtration" (Bloch and Ogus [5]). When the ground field is the complex numbers, étale cohomology can be replaced by ordinary (singular) cohomology. It is known by work of Atiyah and Hirzebruch that torsion cohomology classes are not always carried by algebraic cycles [1] (unlike the situation for $H^2(X, \mathbf{Z})$, whose torsion subgroup is known to come from Chern classes of divisors). Is it possible, however, that all torsion classes die in the complement of a divisor?

References for Lecture 5

[1] M. Atiyah and F. Hirzebruch, Analytic cycles on complex manifolds, *Topology*, **1** (1962) 25–45.

[2] H. Bass and J. Tate, The Milnor ring of a global field, in *Algebraic K-Theory II*, Lecture Notes in Math., no. 342, Springer, Berlin (1973).

[3] S. Bloch, Torsion algebraic cycles and a theorem of Roitman, *Compositio Math.*, **39** (1979), 107–127.

[4] S. Bloch, Torsion algebraic cycles, K_2, and Brauer groups of function fields, *Bull. Amer. Math. Soc.*, **80** (1974), 941–945.

[5] S. Bloch and A. Ogus, Gersten's conjecture and the homology of schemes, *Ann. Sci. École Norm. Sup. (4)*, **7** (1974), 181–212 (1975).

[6] P. Deligne, *Cohomologie étale* (SGA $4\frac{1}{2}$), Lecture Notes in Math., no. 569, Springer, Berlin (1977).

[7] R. Elman and T. Y. Lam, On the quaternion symbol homomorphism $g_F \colon k_2(F) \to B(F)$, in *Algebraic K-Theory II*, Lecture Notes in Math., no. 342, Springer, Berlin (1973).

[8] S. Kleiman, Algebraic cycles and the Weil conjectures, in *Dix exposes sur la cohomologie des schemas*, North Holland, Amsterdam (1968).

[9] S. Lang, *Abelian Varieties*, Interscience Publishers (1959). [Reprinted Springer, 1983.]

[10] J. Tate, Relations between K_2 and Galois cohomology, *Invent. Math.*, **36** (1976), 257–274.

[11] A. A. Roitman, private correspondence.

6

Complements on $H^2(K_2)$

It is possible to study $H^2(X, \mathcal{K}_2)$ infinitesimally and analytically, as well as algebraically. Although one cannot establish at this time any direct relation between the results obtained and the structure of the Chow group, it is plausible that such relations exist. Frequently such formal manipulation suggests conjectures about cycles which can be tested directly.

Suppose for example X is an algebraic surface defined over $\mathbf{C}$, and let $\mathcal{K}_2^{an}$ denote the sheaf for the complex topology on X obtained by sheafifying the presheaf $U \to K_2(\Gamma(U, O_{X^{an}}))$. One can derive a version of the Gersten–Quillen resolution for $\mathcal{K}_2^{an}$. The problem is that the sheaves involved frequently have non-trivial cohomology so there is no obvious map $H^2(X^{an}, \mathcal{K}_2^{an}) \to \mathrm{CH}^2(X)$. There is, however, an obvious map

$$\mathrm{CH}^2(X) \cong H^2(X^{Zar}, \mathcal{K}_2) \to H^2(X^{an}, \mathcal{K}_2^{an}).$$

The easiest case with which to deal is when $P_g = q = 0$. In this case, conjecturally $\mathrm{CH}^2(X) \cong \mathbf{Z}$.

Theorem (6.1) *Let X be a surface as above, and assume $P_g = q = 0$. Then the image of* $\mathrm{CH}^2(X) \to H^2(X^{an}, \mathcal{K}_2^{an})$ *is* $\mathbf{Z}$, *and* $\mathrm{CH}^2(X) \to$ *Image is the degree map.*

Proof The existence of a $d\log$ map $\mathcal{K}_2^{an} \to \Omega^2_{X^{an}}$ leads to

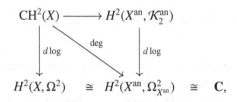

so $\mathrm{Ker}(\mathrm{CH}^2(X) \to H^2(X^{an}, \mathcal{K}_2^{an})) \subset \mathrm{Ker}(\deg)$.

For the other inclusion, let δ be a cycle of degree 0 on X and let $C \subset X$ be a smooth curve with $\operatorname{Supp} \delta \subset C$. From the exact sequence of analytic sheaves on C,

$$0 \to \mathbf{C}^*_{C^{an}} \to O^*_{C^{an}} \to \Omega^1_{C^{an}} \to 0,$$

one gets a class $\eta \in H^1(C^{an}, \mathbf{C}^*)$ mapping to $[\delta] \in H^1(C^{an}, O^*) \cong \operatorname{Pic}(C)$. Let $O_{X^{an}}(\infty C)$ denote that sheaf of analytic functions on X^{an} meromorphic along C. One has sequences

(6.1.1) $0 \to O^*_{X^{an}} \to O_{X^{an}}(\infty C)^* \to j_* \mathbf{Z}_C \to 0, \quad j : C \hookrightarrow X,$

(6.1.2) $0 \to \mathcal{K}^{an}_2 \to \mathcal{K}_2(O_{X^{an}}(\infty C)) \xrightarrow{\text{tame}} j_* O^*_{C^{an}} \to 0,$

and one has a symbol map

$$(6.1.1) \otimes_{\mathbf{Z}} \mathbf{C}^* \to (6.1.2).$$

This gives a commutative diagram

(6.1.3)

$$\eta \in H^1(C^{an}, \mathbf{Z}) \otimes \mathbf{C}^* \xrightarrow{\partial \otimes 1} H^2(X^{an}, O^*_{X^{an}}) \otimes \mathbf{C}^*$$

$$\downarrow{\cong} \qquad\qquad\qquad \downarrow$$

$$H^1(C^{an}, \mathbf{C}^*) \xrightarrow{\partial} H^2(X^{an}, O^*_{X^{an}} \otimes \mathbf{C}^*)$$

$$\downarrow \qquad\qquad\qquad \downarrow$$

$$\operatorname{Pic}(C) \cong H^1(C, O^*_{C^{an}}) \xrightarrow{\partial} H^2(X^{an}, \mathcal{K}^{an}_2).$$

It suffices to note now that the hypotheses $P_g = q = 0$ imply that $H^2(X^{an}, O^*_{X^{an}}) \otimes \mathbf{C}^* = (0)$. Indeed, the exponential sequence

$$0 \to \mathbf{Z}_{X^{an}} \to O_{X^{an}} \to O^*_{X^{an}} \to 0$$

gives

$$H^2(X, O_X) \to H^2(X^{an}, O^*_{X^{an}}) \to H^3(X^{an}, \mathbf{Z}).$$

Since $P_g = q = 0$, the group on the left vanishes and that on the right is finite.   □

The rest of this lecture is devoted to the infinitesimal structure of $H^2(X, \mathcal{K}_2)$. We fix a perfect ground field k which in the case of characteristic zero we assume to be algebraic over $\mathbf{Q}$. (Essentially, when k is not algebraic over the prime field, a cycle on some variety over k must be viewed as the germ of a family of cycles. In the presence of non-trivial derivations of k, the deformations of k-cycles will be more complicated.) Let X be a smooth geometrically connected k-variety. We consider a functor CH_X^2 from the category C of augmented artinian k-algebras to abelian groups defined by

$$\widehat{\mathrm{CH}}_X^2(A) = \mathrm{Ker}(H^2(X_k \times A, \mathcal{K}_2) \xrightarrow[\text{augmentation}]{} H^2(X, \mathcal{K}_2)).$$

Consider first the case $\mathrm{char}\, k = 0$. The basic local result is

Theorem (6.2) *Let R be a local k-algebra, and A an augmented artinian k-algebra with augmentation ideal $\mathfrak{m}$. Put $S = R \otimes_k A$ and $I = R \otimes_k \mathfrak{m}$, and define*

$$K_2(S, I) = \mathrm{Ker}(K_2(S) \to K_2(R)),$$
$$\Omega_{S,I}^1 = \mathrm{Ker}(\Omega_S^1 \to \Omega_R^1),$$

where Ω^1 is the module of absolute Kähler differentials. The universal derivation $d \colon S \to \Omega_S^1$ induces $d \colon I \to \Omega_{S,I}^1$, and we have an isomorphism $K_2(S, I) \cong \Omega_{S,I}^1/dI$.

The proof is purely algebraic and will not be given in these notes. In terms of symbols, the map

$$K_2(S, I) \to \Omega_{S,I}^1/dI$$

is given by

(6.2.1) $\qquad \{1 + \iota, s\} \to \log(1 + \iota)\, \dfrac{ds}{s}, \qquad \iota \in I,\ s \in S^*.$

One has an exact sequence ($k' =$ algebraic closure of k in R)

(6.2.2) $\qquad k' \otimes_k \mathfrak{m} \xrightarrow{1 \otimes d} R \otimes_k \Omega_{A,\mathfrak{m}}^1 \to \Omega_{S,I}^1/dI \to (\Omega_R^1/dR) \otimes_k \mathfrak{m} \to 0,$

which can be used to analyze the structure of $K_2(S, I)$.

Thinking of R as the local ring at a variable point on X, we may pass to Zariski sheaves, getting an exact sequence of sheaves of k-vector spaces

(6.2.3)

$$0 \to k \otimes_k d\mathfrak{m} \to O_X \otimes_k \Omega_A^1 \to \mathcal{K}_{2_{X \times A, X \times \mathfrak{m}}} \longrightarrow (\Omega_X^1/dO_X) \otimes_k \mathfrak{m} \longrightarrow 0$$

$$(\Omega_X^1 \otimes_k \mathfrak{m} \,/\, dO_X \otimes_k \mathfrak{m}').$$

Here $\mathfrak{m}' = \mathrm{Ker}(d\colon \mathfrak{m} \to \Omega^1_A)$. Hodge theory gives an exact sequence when X is complete:

$$0 \to H^*(X, \Omega^1_X) \to H^*(X, \Omega^1_A/dO_X) \to H^{*+1}(X, O_X) \to 0.$$

In this case the long exact sequence of cohomology associated to (6.2.3) looks like

(6.2.4)

$$0 \to H^2(X, O_X) \otimes (\Omega^1_A/d\mathfrak{m}) \to \widehat{\mathrm{CH}}^2_X(A)$$

$$\to H^2(X, \Omega^1_X/dO_X) \otimes \mathfrak{m} \longrightarrow H^3(X, O_X) \otimes \Omega^1_A$$

$$\quad \searrow \qquad\qquad\qquad \Big\uparrow {\scriptstyle 1\otimes d}$$

$$H^3(X, O_X) \otimes \mathfrak{m}.$$

Remark (6.2.5) (i) When $H^3(X, O_X) = (0)$ (e.g. $\dim X = 2$) or when $d\colon \mathfrak{m} \hookrightarrow \Omega^1_A$, one gets

(6.2.6) $\quad 0 \to H^2(X, O_X) \otimes (\Omega^1_A/d\mathfrak{m}) \to \widehat{\mathrm{CH}}^2_X(A) \to H^2(X, \Omega^1_X) \otimes \mathfrak{m} \to 0.$

There do exist, however, artinian k-algebras for which d is not injective, so in general the right-hand term is more complicated.

(ii) A functor $F\colon C \to$ (sets) is said to be *pro-representable* if there exists a complete local k-algebra Λ and an isomorphism of functors

$$F(\cdot) \cong \mathrm{Morph}_{\text{local } k\text{-algebra}}(\Lambda, \cdot).$$

The functor $A \to H^2(X, \Omega^1_X) \otimes \mathfrak{m}$ is pro-represented by the completion at 0 of the symmetric algebra on the dual vector space to $H^2(X, \Omega^1_X)$. When $H^3(X, O_X) = (0)$, (6.2.6) displays $\widehat{\mathrm{CH}}^2_X$ as a pro-representable quotient and a non pro-representable kernel. If $H^2(X, O_X) = H^3(X, O_X) = (0)$ we see that $\widehat{\mathrm{CH}}^2_X$ is itself pro-representable, a result which should be compared with Lecture 1. For example, when X is a surface, we find $\widehat{\mathrm{CH}}^2_X$ isomorphic to the formal group at the origin of the Albanese if and only if $H^2(X, O_X) = (0)$.

Example (6.2.7) Let $A = k[\varepsilon, \delta](\varepsilon^2, \delta^2, \varepsilon\delta)$. Consider deformations of the trivial divisor

$$D_\varepsilon \in \mathrm{Ker}(H^1(X \times k[\varepsilon], O^*) \to H^1(X, O^*)) \cong H^1(X, O_X),$$

$$E_\delta \in \mathrm{Ker}(H^1(X \times k[\delta], O^*) \to H^1(X, O^*)) \cong H^1(X, O_X),$$

and consider the product

$$D_\varepsilon \cdot E_\delta \in \mathrm{Ker}(H^2(X \times A, \mathcal{K}_2) \to H^2(X \times k[\varepsilon], \mathcal{K}_2) \oplus H^2(X \times k[\delta], \mathcal{K}_2))$$
$$\cong H^2(X, O_X).$$

If D_ε and E_δ are represented by 1-cocycles $1 + d_{ij}\,\varepsilon$ and $1 + e_{ij}\,\delta$ respectively, the intersection is represented by $\{1 + d_{ij}\,\varepsilon, 1 + e_{jk}\,\delta\}$. Viewed as an element in $H^2(X, O_X) \otimes_k (\Omega^1_A / d\mathfrak{m})$, we find $D_\varepsilon \cdot E_\delta$ represented by $\{d_{ij}\,e_{jk}\} \otimes \varepsilon\, d\delta$. Identifying $\Omega^1_A / d\mathfrak{m} \cong k$ with generator $\varepsilon\, d\delta$, it follows that intersection of cycles is given by cup product $H^1(O_X) \otimes H^1(O_X) \to H^2(O_X)$. Thus, the non pro-representable kernel in (6.2.6) does play a role in the cycle theory.

I want finally to discuss the work of Stienstra on $\widehat{\mathrm{CH}}_X^2$ when X is defined over a perfect field k of characteristic $p \neq 0$. If R is a ring, the big Witt ring of R, big $W(R)$, is defined additively by

$$\text{big } W(R)^+ = (1 + Rt[[t]])^* = \text{group of power series in } R$$
$$\text{with constant term } 1.$$

The multiplication is determined by functoriality in R together with the requirement $(1 - at)^{-1} * (1 - bt)^{-1} = (1 - abt)^{-1}$. When $pR = 0$, big $W(R)$ splits into a product of rings $W(R)$,

$$W(R) = \left\{ E(a_0 t) E(a_1 t^p) E(a_2 t^{p^2}) \cdots \right\},$$

where

$$E(t) = \exp\left(t + \frac{t^p}{p} + \frac{t^{p^2}}{p^2} + \cdots \right) = \text{Artin–Hasse exponential.}$$

For example, k being perfect one finds that $W(k)$ is a complete characteristic zero discrete valuation ring with maximal ideal $pW(k)$ and residue field k.

$W(R)$ has a ring endomorphism F induced by the Frobenius $r \to r^p$ on R and an additive endomorphism V given by $P(t) \to P(t^p)$, P being a power series in t. One has $FV = VF = p$. The Frobenius on $W(k)$ is an automorphism and is usually denoted σ. The relation $V(Fx \cdot y) = xV(y)$ holds in $W(R)$, so $W(R)$ is a module for the *Dieudonné ring* $D = W(k)[F, V]$. Relations in D are $FV = VF = p$, $Fw = w^\sigma F$, $Vw = w^{\sigma-1}V$ with $w \in W(k)$. An important variant on these ideas arises by defining

$$\text{big } \widehat{W}(R) = \mathrm{Ker}(R[T]^* \underset{T \mapsto 1}{\to} R^*).$$

Big $\widehat{W}(R)$ is a module for big $W(R)$ and inherits the same splitting in characteristic p:

$$\widehat{W}(R) = \{ E(a_0 t) \cdots E(a_n t^{p^n}) \mid a_i \text{ nilpotent} \}.$$

We view $\widehat{W}$ as a functor

$$\text{(artinian } k\text{-algebras)} \xrightarrow{\widehat{W}} \text{(left } D\text{-modules)}.$$

(6.3.1)   Let M be a left D-module with the following properties:

   (i) M is V-adically complete and separated,
   (ii) M has no V-torsion,
   (iii) M/VM is a $W(k)$-module of finite length.

For example, $M = D/D \cdot (F^n - V^m)$ or $M = k[[V]]$. We view M as a right D-module via

$$mw = wm, \quad mV = Fm, \quad mF = Vm.$$

The tensor product $M \otimes_D \widehat{W}$ then becomes a functor

$$M \otimes_D \widehat{W} : \text{(artinian } k\text{-algebras)} \to \text{(abelian groups)},$$

which is pro-representable, formally smooth (i.e. a surjection $A \to B$ of artinian k-algebras induces a surjection $M \otimes_D \widehat{W}(A) \to M \otimes_D \widehat{W}(B)$) and has finite-dimensional "tangent space"

$$M \otimes_D \widehat{W}(k[\varepsilon]/(\varepsilon^2)) \cong M/VM.$$

Such a functor is called a (smooth) formal group, and it turns out that any smooth formal group can be represented as $M \otimes_D \widehat{W}$ for some M.

Examples (6.3.2)   (i) $M = W(k)$, $F = \sigma$, and $V = p\sigma^{-1}$. Then

$$W(k) \otimes_D \widehat{W} \cong \widehat{W}/(1 - V)\widehat{W} \cong \widehat{\mathbf{G}}_{\mathrm{m}},$$

where $\widehat{\mathbf{G}}_{\mathrm{m}}$ is the formal multiplicative group and the map $\widehat{W} \to \widehat{\mathbf{G}}_{\mathrm{m}}$ is defined by

$$E(a_0 t) \cdots E(a_n t^{p^n}) \to E(a_0) \cdots E(a_n).$$

   (ii) $M = k[[V]]$ and $F = 0$. Then $M \otimes_D \widehat{W} \cong \widehat{W}/V\widehat{W} \cong \widehat{\mathbf{G}}_{\mathrm{a}}$, the formal additive group.

As seen in characteristic zero, $\widehat{\mathrm{CH}}_X^2$ is not in general pro-representable. This means that $\widehat{W}$ and related functors are inadequate for its description. The trick is to think of $\widehat{W}$ as being associated to units in a polynomial ring, that is to K_1. This suggests considering "formal curves" on K_2:

$$\widehat{C}K_2 : \text{(augmented artinian } k\text{-algebras)} \longrightarrow \text{(abelian groups)}$$

$$A \longrightarrow \mathrm{Ker}\,(K_2(A[X]) \xrightarrow{X \mapsto 0} K_2(A)).$$

One can also define for any k-algebra R, curves of length n on K_2:

$$C_n K_2(R) = \mathrm{Ker}\,(K_2(R[X]/X^{n+1}) \to K_2(R)),$$

and curves on K_2:

$$CK_2(R) = \varprojlim_n C_n K_2(R).$$

There are bilinear pairings

$$\mathrm{big}\ W(R) \times \widehat{C}K_2(A) \to K_2(R \otimes_k A),$$
$$(1 - rT)^{-1} \times \alpha(X) \mapsto \alpha(r)$$

and

$$CK_2(R) \times \mathrm{big}\ \widehat{W}(A) \to K_2(R \otimes_k A),$$
$$\beta(x) \times (1 - aT)^{-1} \mapsto \beta(a).$$

(This makes sense because a is nilpotent.)

Also, when char $k = p$, there are splittings analogous to splittings on big W and big $\widehat{W}$: $\widehat{C}K_2(A) = \prod_\infty T\widehat{C}K_2(A)$ (*formal typical curves*) and $CK_2(R) = \prod_\infty TCK_2(R)$ (*typical curves*). TCK_2 and $T\widehat{C}K_2$ are D-modules, and the above pairings give maps

$$\Phi\colon W(R) \otimes_D T\widehat{C}K_2(A) \to K_2(R \otimes_k A, R \otimes_k M)$$
$$= \mathrm{Ker}\,(K_2(R \otimes_k A) \to K_2(R)),$$
$$\Psi\colon TCK_2(R) \otimes_D \widehat{W}(A) \to K_2(R \otimes_k A, R \otimes_k M).$$

This suggests that the role of Ω^1 in characteristic p is played by TCK_2. Pursuing this analogy, there is a map

$$d\colon W(R) \to TCK_2(R)$$

defined by $d(P(t)) = \{P(t), t\}$, where $P(t)$ is a monic power series and $\{\cdot, \cdot\}$ is the Steinberg symbol. (This symbol can be shown to make sense even though t is not a unit.) Stienstra writes $\overrightarrow{\partial W(R)}$ for the D-submodule generated by $d(W(R))$ in $TCK_2(R)$. His main local result is

Theorem (6.3.3) *Let R be a regular local k-algebra and A an augmented artinian local k-algebra. Then the composite*

$$TCK_2(R) \otimes_D \widehat{W}(A) \overset{\Psi}{\to} K_2(R \otimes A, R \otimes \mathfrak{m}) \to \mathrm{Coker}\,\Phi$$

is trivial on $\partial W(R) \otimes_D \widehat{W}(A)$, and there is an induced exact sequence

$$0 \to W(R) \otimes_D T\widehat{C}K_2(A) \to K_2(R \otimes_k A, R \otimes_k \mathfrak{m})$$
$$\to (TCK_2(R)/\partial W(R)) \otimes_D \widehat{W}(A) \to 0.$$

This whole process can be sheafified. One works with presheaves

$$\text{``}\varprojlim\text{''} W_n(O_X), \qquad \text{``}\varprojlim\text{''}(TC_nK_2(O_X)/\partial W(O_X))$$

and cohomology groups

$$H^*(X, W) := \varprojlim_n H^*(X, W_n(O_X)),$$

$$H^*(X, TCK_2/\partial W) := \varprojlim_n H^*(X, TC_nK_2(O)/\partial W(O)).$$

The former was first studied by Serre [8]. It is known to satisfy (6.3.1) (i)–(iii), and the formal groups $H^*(X, W) \otimes_D \widehat{W}$ have been studied by Artin and Mazur. The groups $H^*(X, TCK_2/\partial W)$ may according to Stienstra have V-torsion and so not satisfy (6.3.1)(ii). The main geometric consequence of (6.3.3) is

Theorem (6.3.4) *Let X be a smooth complete surface over a perfect field of characteristic $p > 0$. Then there is an exact sequence for any augmented local artinian k-algebra A*

$$\cdots \to H^1(X, (TCK_2/\partial W) \otimes_D \widehat{W}(A)) \xrightarrow{\delta} H^2(X, W) \otimes_D T\widehat{C}K_2(A)$$

$$\to \widehat{CH}^2_X(A) \to H^2(X, TCK_2/\partial W) \otimes_D \widehat{W}(A) \to 0.$$

If $H^2(X, W)$ has no p-torsion, the map δ is zero.

Remarks (6.3.5) (i) The point of taking X to be a surface is that H^2 is right exact, so the $\otimes_D$ can be brought outside the cohomology.

(ii) It seems certain that $H^2(X, TCK_2/\partial W)/(V\text{-torsion})$ is the Dieudonné module of the Albanese variety of X, although no one has written down a proof. In the absence of V-torsion, this would make $H^2(X, TCK_2/\partial W) \otimes_D \widehat{W}$ the formal group at the origin of the Albanese. It is certainly true under mild hypotheses that $H^2(X, TCK_2/\partial W)$ is "the part" of crystalline $H^3(X)$ with slopes in $[1, 2)$, but one doesn't know precisely the correct hypotheses to impose at this point.

(iii) The moral is that $\widehat{CH}^2_X$ is controlled by the D-modules $H^2(X, W)$ and $H^2(X, TCK_2/\partial W)$, and these two (playing the role of $H^2(X, O_X)$ and $H^2(X, \Omega^1_X)$ in characteristic zero) have a tendency to be Dieudonné modules, that is to satisfy (6.3.1)(i)–(iii).

The non pro-representable part of $\widehat{CH}_x^2$ differs from a formal group in that $\otimes_D \widehat{W}$ is replaced by $\otimes_D T\widehat{C}K_2$. It would be of interest to study algebraically functors of the form $M \otimes_D T\widehat{C}K_2$. Can one give an intrinsic characterization of such a functor? Can one understand the link between the algebraic properties of these functors and the geometric properties of the cycles?

References for Lecture 6

[1] M. Artin and B. Mazur, Formal groups arising from algebraic varieties, *Ann. Sci. École Norm. Sup. (4)*, **10** (1977), 87–132.

[2] S. Bloch, K_2 of Artinian **Q**-algebras, with applications to algebraic cycles, *Comm. Algebra*, **3** (1975), 405–428.

[3] S. Bloch, Algebraic K-theory and crystalline cohomology, *Inst. Hautes Études Sci. Publ. Math.*, no. 47 (1977), 187–268 (1978).

[4] S. Bloch, Some formulas pertaining to the K-theory of commutative group schemes, *J. Algebra*, **53** (1978), 304–326.

[5] S. Bloch, Applications of the dilogarithm function in algebraic K-theory and algebraic geometry, pp. 103–114 in *Proceedings of the International Symposium on Algebraic Geometry (Kyoto Univ., Kyoto, 1977)*, Kinokuniya Book Store, Tokyo (1978).

[6] J. Stienstra, The formal completion of the second Chow group: a K-theoretic approach, pp. 149–168 in *Journée de Géométrie Algébrique de Rennes (Rennes, 1978), Vol. II, Asterisque*, **64** (1979).

[7] J. Stienstra, Deformations of the second Chow group, Thesis, Utrecht (1978).

[8] J.-P. Serre, Sur la topologie des variétés algebriques en caractéristique p, pp. 24–53 in *Symposium Internacional de Topologia Algebraica*, Universidad Nacional Autónoma de Mexico and UNESCO, Mexico City (1958).

Diophantine questions

A smooth projective algebraic surface X defined over a field k is said to be *rational* if its function field becomes rational after extension to the algebraic closure $\bar{k}$ of k, that is $\bar{k}(X) \cong \bar{k}(t_1, t_2)$ with t_1, t_2 independent transcendentals. If $k(X) \cong k(t_1, t_2)$ we say X is *k-rational*. The Chow group $CH_0(X)$ is defined as usual, either geometrically

$$CH_0(X) = \text{Coker}\Big(\coprod_{X^1} k(x)^* \to \coprod_{X^2} \mathbf{Z} \Big)$$

or algebraically by $H^2(X, \mathcal{K}_2)$. The degree map $\deg : CH_0(X) \to \mathbf{Z}$ is a bit tricky. It can be defined either as the composition

$$CH_0(X) \to CH_0(\bar{X}) \xrightarrow{\deg} \mathbf{Z}, \quad \text{where } \bar{X} = X \times_{\text{Spec}\,k} \text{Spec}\,\bar{k},$$

or directly by assigning to $(x), x \in X^2, \deg(x) = [k(x) : k]$. Let $A_0(X) \subset CH_0(X)$ be the kernel of deg.

Proposition (7.1) *If X is k-rational, then $A_0(X) = (0)$.*

Proof Let $f : \mathbf{P}_k^2 - \to X$ be a birational map, defined over k, and let Γ be its graph. From the geometric definition of CH_0 one sees easily that a zero-cycle can be moved off of any given proper closed subscheme of X or $\mathbf{P}^2$. The correspondences

$$CH_0(X) \xrightarrow{\Gamma_*^t} CH_0(\mathbf{P}^2) \xrightarrow{\Gamma_*} CH_0(X)$$

are therefore inverse isomorphisms. One knows $CH_0(\mathbf{P}^2) \cong \mathbf{Z}$ (exercise). □

The same argument shows, of course, that $A_0(X)$ is a k-birational invariant. To simplify exposition we will tend to assume k is either a local or a global field. By a technique of Manin involving Brauer groups, we will show $A_0(X)$

is not in general zero for X a rational surface. We then introduce the Néron–Severi torus, $T = T_X$, which is the k-torus with character group the $\mathrm{Gal}(\bar{k}/k)$-module

$$N = \mathrm{NS}(\bar{X}) = \mathrm{CH}^1(\bar{X}).$$

The rationality of X implies that N is a finitely generated free abelian group. By definition,

$$T(\bar{k}) = \mathrm{Hom}_{\mathbf{Z}}(N, \bar{k}^*)$$

with the natural Galois action. This torus has been studied by Colliot-Thélène and Sansuc [2]. They consider a map (as usual $H^*(k, M) = H^*_{\mathrm{Gal}}(\mathrm{Gal}(\bar{k}/k), M)$)

$$\Phi \colon A_0(X) \to H^1(k, T(\bar{k}))$$

and show (assuming k to be a local or global field) that Image(Φ) is finite.

Using K-theory, we construct an exact sequence ($\bar{F} = \bar{k}(X)$)

$$(7.2) \quad H^0(k, T(\bar{k})) \longrightarrow H^1(k, K_2(\bar{F})/K_2(\bar{k})) \overset{\mu}{\longrightarrow} A_0(X)$$
$$\overset{\Phi}{\longrightarrow} H^1(k, T(\bar{k})) \longrightarrow H^2(k, K_2(\bar{F})/K_2(\bar{k})).$$

When X has a rational pencil of genus-zero curves ($X = $ *conic bundle surface*) we will show for k local or global that $H^1(k, K_2(\bar{F})/K_2(\bar{k}))$ is a finite 2-group, so $A_0(X)$ is finite in this case. The heart of the argument is an analog of the Eichler norm theorem describing the image of the reduced norm

$$\mathrm{Nrd} \colon A^* \to k(t)^*,$$

when A is a quaternion algebra over the rational function field $k(t)$ (for k local or global). As a corollary, we show that if X is a smooth cubic surface (i.e. $X \subset \mathbf{P}^3_k$ is cut out by a homogeneous function of degree 3), then the only possible infinite torsion in $A_0(X)$ is 3-torsion.

Conjecture (7.3) $A_0(X)$ *is finite for any rational surface over a local or global field.*

Question (7.4) As the reader will see, the structure of $A_0(X)$ is somehow linked to the places of k where X has bad reduction. Does the order of

$$\mathrm{Image}(A_0(X) \to A_0(X_{k_v}))$$

for the various completions of k have anything to do with the local factors at bad places in the zeta function of X?

To convince ourselves there is some interest in the subject, we must show that $A_0(X)$ is not in general zero for a rational surface. The idea is due to Manin, but I want to give a very "motivic" (some might say very pedantic) exposition. Let C be the category of finite separable extension fields of k. (We avoid the temptation to work with the larger category of all k-algebras because of possible technical problems.) We consider various covariant functors on C

$$X \colon C \to (\text{sets}), \quad k' \to X(k') \quad (\text{functor of points}),$$
$$\text{Br} \colon C \to (\text{abelian groups}), \quad k' \to \text{Br}(k') = H^2(k', \bar{k}^*),$$
$$A_0(X) \subset \text{CH}_0(X) \colon C \to (\text{abelian groups}), \quad k' \to A_0(X_{k'}) \subset \text{CH}_0(X_{k'}).$$

There is a natural morphism of functors $X \times X \to A_0$, where $(x_1, x_2) \mapsto (x_1) - (x_2)$. An element $a \in H^2_{\text{et}}(X, \mathbf{G}_m)/H^2(k, \bar{k}^*)$ defines a morphism of functors $\tilde{a} \colon X \times X \to \text{Br}$,

$$\tilde{a}(x_1, x_2) = a(x_1) \cdot a(x_2)^{-1} \in \text{Br}(k'), \qquad x_1, x_2 \in X(k').$$

Manin's observation is that there exists a morphism $A_0(X) \xrightarrow{\psi_a} \text{Br}$ making the diagram

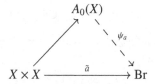

commute. In particular, if $\tilde{a}_k \colon X(k) \times X(k) \to \text{Br}(k)$ is not zero, it follows that the group $A_0(X_k)$ isn't either.

To define ψ_a, we associate to a zero-cycle $\sum n_i (x_i)$ defined over k'

$$\psi_a \left(\sum n_i (x_i) \right) = \prod_i \text{cor}_{k'(x_i)/k'} (a(x_i))^{n_i} \in \text{Br}(k').$$

Given a correspondence defined over k'

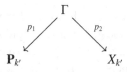

we have (using the existence of transfer for the Brauer group)

$$\psi_a(p_{2*}p_1^*((0) - (\infty))) = \psi_{p_2^*(a)}(p_1^*((0) - (\infty)))$$
$$= \psi_{p_{1*}p_2^*(a)}((0) - (\infty)).$$

But $p_{1*}p_2^*(a) \in \mathrm{Br}(\mathbf{P}_{k'}^1)/\mathrm{Br}(k') = (0)$, so ψ_a is trivial on cycles rationally equivalent to zero, as desired.

Lemma (7.5) *Let X be a rational surface. Then*

$$\mathrm{Br}(X)/\mathrm{Br}(k) \cong H^1(\mathrm{Gal}(\bar{k}/k), \mathrm{Pic}(\bar{X})).$$

Proof The Brauer group of a surface is a birational invariant, so

$$H_{\mathrm{et}}^2(\bar{X}, \mathbf{G}_m) \cong H_{\mathrm{et}}^2(\mathbf{P}_{\bar{k}}^2, \mathbf{G}_m) = (0)$$

(Grothendieck [11]). The Hochshild–Serre spectral sequence

$$E_2^{p,q} = H^p(\mathrm{Gal}(\bar{k}/k), H_{\mathrm{et}}^q(\bar{X}, \mathbf{G}_m)) \Rightarrow H_{\mathrm{et}}^{p+q}(X, \mathbf{G}_m)$$

gives rise, therefore, to an exact sequence

$$\mathrm{Br}(k) \to \mathrm{Br}(X) \to H^1(\mathrm{Gal}(\bar{k}/k), H^1(\bar{X}, \mathbf{G}_m)) \to H^3(\mathrm{Gal}(\bar{k}/k), \bar{k}^*).$$

The right-hand term vanishes for k local or global. □

Remarks on calculating $H^1(\mathrm{Gal}(\bar{k}/k), \mathrm{Pic}(\bar{X}))$ If $k \subset K$ is a finite extension such that $\mathrm{Gal}(\bar{k}/K)$ acts trivially on $\mathrm{Pic}(\bar{X})$, one checks easily

$$H^1(\mathrm{Gal}(\bar{k}/k), \mathrm{Pic}(\bar{X})) \cong H^1(\mathrm{Gal}(K/k), \mathrm{Pic}(\bar{X})).$$

Suppose now F is a finitely generated free abelian group of divisors on X_K such that $F \to \mathrm{Pic}(X_K)$. Let

$$L = \{f \in K(X)^* \mid (f) \in F\}$$

and consider the exact sequence

$$0 \to L/K^* \to F \to \mathrm{Pic}(X_K) \to 0.$$

One has $H^1(\mathrm{Gal}(K/k), F) = (0)$ (F is a permutation module) whence an exact sequence

$$0 \to \mathrm{Br}(X)/\mathrm{Br}(k) \xrightarrow{\partial} H^2(\mathrm{Gal}(K/k), L/F^*) \to H^2(\mathrm{Gal}(K/k), F).$$

In simple situations, Manin actually writes down 2-cochains $f_{s,t}$ with values in $L \subset K(X)^*$ such that the $f_{s,t} \mod(K^*)$ represent the image of $\mathrm{Br}(X)$ in $H^2(\mathrm{Gal}(K/k), L/K^*)$. Given a zero-cycle $\sum n_i(x_i)$ defined over k and supported in the complement of the support of L, we have $\psi_a(\sum n_i(x_i))$ represented by the 2-cocycle

$$\prod_i N_{K(x_i)/K}(f_{s,t}(x_i))^{n_i}$$

with values in K^*. Here $a \in \mathrm{Br}(X)/\mathrm{Br}(k)$ with ∂a represented by $f_{s,t}$.

When K/k is cyclic, one can use the periodicity of Galois cohomology to identify

$$H^2(\mathrm{Gal}(K/k), L/K^*) \cong L^{\mathrm{Gal}(K/k)}/k^* N_{K(X)/k(X)}(L),$$

$$\mathrm{Br}(k) \supset H^2(\mathrm{Gal}(K/k), K^*) \cong k^*/N_{K/k}(K^*).$$

In this case one can write down $f \in L^{\mathrm{Gal}(K/k)} \subset k(X)^*$ such that $\sum n_i(x_i)$ is not rationally equivalent to 0 if

$$\prod_i N_{k(x_i)/k} f(x_i)^{n_i} \notin NK^* \subset k^*.$$

Example (7.6) Châtelet surfaces are defined in affine coordinates by $t_1^2 - a t_2^2 = (x - a_1)(x - a_2)(x - a_3)$, for $a, a_i \in k^*$, $a_i \neq a_j$ (Châtelet [1], Manin [5]). In this case we can take $K = k(\sqrt{a})$. Manin shows that given k-points p_1, p_2 on the surface such that the functions

$$f_1 = \frac{x - a_1}{x - a_3}, \qquad f_2 = \frac{x - a_2}{x - a_3}$$

are defined and invertible at p_1 and p_2, if at least one of the ratios

$$\frac{f_i(p_1)}{f_i(p_2)} \notin N_{K/k}(K^*), \quad i = 1, 2,$$

then p_1 is not rationally equivalent to p_2. Take for example

$$t_1^2 - 5t_2^2 = x(x - 1)(x - 7), \quad k = \mathbf{Q}.$$

For each value of $x \neq 0, 1, 7$ we obtain a conic curve, the rationality of which (over $\mathbf{Q}$) is determined by the Hilbert symbol

$$(5, x(x - 1)(x - 7)).$$

To obtain rational fibres take $x = -1$,

$$(5, (-1)(-2)(-8)) = (5, -1) = 1 \quad (-1 \equiv 2^2 \bmod 5)$$

and $x = 31$,

$$(5, 31 \cdot 30 \cdot 24) \equiv (5, 31 \cdot 5) = 1$$

($5 \equiv 6^2 \bmod 31$ and $31 \equiv 1 \bmod 5$; also $(5, 5) = (5, -1) = 1$). Let p_1 be a rational point in the curve over $x = -1$, p_2 a rational point over $x = 31$:

$$f_1(p_1) = \frac{-1}{-8} \equiv 2 \quad \bmod N_{K/k} k^*,$$

$$f_1(p_2) = \frac{31}{24} \equiv \frac{31}{6} \quad \bmod N_{K/k} k^*.$$

To show p_1 and p_2 not rationally equivalent, we may at this point pass to a larger field, working with the 5-adic completions $k_5 = \mathbf{Q}_5$ and $K_5 = \mathbf{Q}_5(\sqrt{5})$. The field K_5 is tamely ramified over k_5, so local class field theory implies

$$N_{K_5/k_5} K_5^* = 5^{\mathbf{Z}} \cdot (1 + 5O_{k_5})^* \cdot \mu_{k_5}^2, \qquad \mu_{k_5} \subset k_5^* \text{ roots of 1.}$$

Since 2 is not a square mod 5, we get $f_1(p_1) \notin N_{K_5/k_5} K_5^*$. On the other hand, $\frac{31}{6} \equiv 1$ mod 5, so $\frac{31}{6} \in N_{K_5/k_5} K_5^*$. We conclude that p_1 is not rationally equivalent to p_2, even over $\mathbf{Q}_5$.

Our objective now is to construct the fundamental exact sequence (7.2).

Proposition (7.7) *Let X be a smooth surface defined over k, and let Y be obtained from X by blowing up a k-point. Then $\Gamma(X, \mathcal{K}_2) = \Gamma(Y, \mathcal{K}_2)$ and $H^1(X, \mathcal{K}_2) \oplus k^* = H^1(Y, \mathcal{K}_2)$.*

Proof Let $F = k(X)$ and let e be the generic point of the exceptional divisor E on Y. Consider the diagram

(7.7.1)

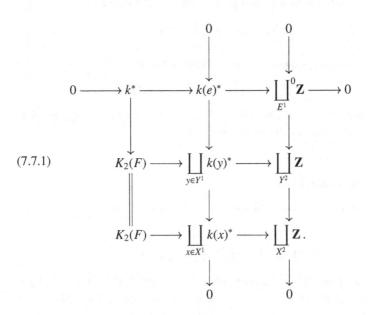

(As usual, the superscript 0 indicates that the right-hand column is in degree 0.) The middle and right-hand columns and the upper row are exact. The middle and lower rows are complexes whose cohomology is identified with the cohomology of $\mathcal{K}_2$ on Y (resp. X) (cf. Lecture 4). These complexes are covariant functorial for proper maps. Using the contravariant functoriality of $\Gamma(\cdot, \mathcal{K}_2)$,

we get a commutative diagram

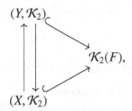

whence $\Gamma(X, \mathcal{K}_2) \cong \Gamma(Y, \mathcal{K}_2)$. The assertions about H^1 follow from diagram chasing on (7.7.1). □

Definition (7.8)   A rational surface X defined over k will be said to be *split* over an extension field k' if there exists a diagram

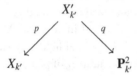

such that the arrows p and q arise from a succession of blowings up of k'-points.

It is known that there always exists k' finite over k splitting X.

Proposition (7.9)   *Let X be a k-rational surface split over k. Then* $\mathrm{Pic}(X) \cong \mathrm{Pic}(\bar{X})$ *and* $H^1(X, \mathcal{K}_2) \cong \mathrm{Pic}(X) \otimes_{\mathbf{Z}} k^*$.

Proof   The assertions of the proposition are true for $X = \mathbf{P}^2$. (For computations of $H^p(\mathbf{P}^n, \mathcal{K}_q)$ see Gillet [14], Sherman [15].) One now applies (7.7) noting that $\mathrm{Pic}(Y) \cong \mathrm{Pic}(X) \oplus \mathbf{Z}$, and checking that the extra factor of k^* in $H^1(Y, \mathcal{K}_2)$ comes from the map

$$\mathrm{Pic}(Y) \otimes k^* = H^1(Y, \mathbf{G}_m) \otimes k^* \to H^1(Y, \mathcal{K}_2)$$

deduced from the Steinberg symbol $\mathbf{G}_m \otimes k^* \to K_2$. □

Now let X be a rational surface defined over k, and let k' be a Galois extension of k splitting X. Note in particular that X will be k'-rational, so $A_0(X_{k'}) = (0)$ by (7.1). Write $X' = X_{k'}$, $F' = k'(X)$, and $N = \mathrm{Pic}(X')$. We get exact sequences (defining Z)

$$(7.10) \qquad 0 \to Z \to \coprod_{x' \in X'^1} k'(x')^* \to \coprod_{X'^2}^{0} \mathbf{Z} \to 0,$$

$$0 \to K_2(F')/K_2(k') \to Z \to N \otimes k'^* \to 0.$$

One has

$$H^*(k'/k, \textstyle\coprod_{X'^1} k'(x')^*) \cong \coprod_{x\in X^1} H^*(k'/k_x, k'(x)^*)$$

by Shapiro's lemma, where $k_x = k' \cap k(x)$. Also

$$\Gamma(k'/k, \textstyle\coprod_{X'^2} {}^0\mathbf{Z}) \cong \coprod_{X^2} {}^0\mathbf{Z}.$$

It follows from the first sequence in equation (7.10) and Hilbert 90 that $A_0(X) \cong H^1(k'/k, Z)$. Substituting in the second sequence yields

$$(7.11) \quad \Gamma(k'/k, N\otimes k') \rightarrow H^1(k'/k, K_2(F')/K_2(k')) \rightarrow A_0(X)$$
$$\rightarrow H^1(k'/k, N\otimes k') \rightarrow H^2(k'/k, K_2(F')/K_2(k'))\,.$$

Finally, (7.11) may be identified with (7.2) by observing that the intersection pairing on N is perfect, so $N\otimes k'^* \cong \text{Hom}(N, k'^*)$ as Galois modules.

Let v be a non-archimedean place of the field k. A surface X defined over k is said to have *good reduction* at v if there exists $\tilde{X} \rightarrow \text{Spec}(O_v)$ smooth and projective with $\tilde{X}_{k_v} \cong X_{k_v}$. For k a global field, any smooth k-surface will have good reduction at all but a finite number of places.

Theorem (7.12) *Let X be a rational surface defined over a local or global field k.*

(i) *If X has good reduction at a non-archimedean place v of k, then the map $\Phi_v : A_0(X_{k_v}) \rightarrow H^1(\bar{k}_v/k_v, T(\bar{k}_v))$ deduced from (7.2) is zero.*

(ii) *Quite generally, the image of $\Phi : A_0(X) \rightarrow H^1(\bar{k}/k, T(\bar{k}))$ is finite.*

Proof Local duality (Serre [13]) implies that $H^1(\bar{k}/k, T(\bar{k}))$ is finite for k a local field. Global duality (Nakayama [12]) says for k global, the kernel of $H^1(\bar{k}/k, T(\bar{k})) \rightarrow \prod_v H^1(\bar{k}_v/k_v, T(\bar{k}_v))$ is finite. These results show that (i) $\Rightarrow$ (ii), so it suffices to prove (i).

Lemma (7.12.1) *Let R be a complete discrete valuation ring with residue field $\mathfrak{f}$ and quotient field k. Let $f : \tilde{X} \rightarrow \text{Spec}\,R$ be smooth, projective, geometrically irreducible, and assume the closed fibre X_0 is a rational surface split over $\mathfrak{f}$. Then the generic fibre X is rational and split over k.*

Proof An $\mathfrak{f}$-valued point of X_0 lifts by smoothness to an R-valued point of $\tilde{X}$, so a morphism $X_0(1) \rightarrow X_0$ obtained by blowing up an $\mathfrak{f}$-point lifts to a diagram

$$
\begin{array}{ccc}
X_0(1) & \subset & \tilde{X}(1) \\
\downarrow & & \downarrow {\scriptstyle \text{blow up an } R\text{-point}} \\
X_0 & \subset & \tilde{X}.
\end{array}
$$

We may thus build a figure

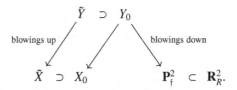

Let $E_0 \subset Y_0$ be an exceptional curve of the first kind. There exists a unique family of exceptional curves of the first kind $\tilde{E} \subset \tilde{Y}$ lifting E_0. Indeed, the normal bundle of $E_0 \cong \mathbf{P}^1$ in Y_0 is $O(-1)$, so both zeroeth and first cohomology groups of the normal bundle vanish, implying uniqueness and existence of liftings.

We may now blow down $\tilde{E}$ on $\tilde{Y}$ in much the same way as E_0 on Y_0. Iterating, we get a diagram

$$\begin{array}{ccc} Y_0 & \subset & \tilde{Y} \\ \downarrow & & \downarrow \\ \mathbf{P}_{\mathfrak{f}}^2 & \subset & \tilde{Z} \end{array}$$

where $\tilde{Z}$ is smooth and projective over $\operatorname{Spec} R$ with closed fibre $\mathbf{P}_{\mathfrak{f}}^2$. Since $\mathbf{P}_{\mathfrak{f}}^2$ is rigid, it follows that $\tilde{Z} \cong \mathbf{P}_R^2$. Passing now to the generic fibres we get $X_k \leftarrow Y_k \to Z_k \cong \mathbf{P}_k^2$, both morphisms being obtained by successive blowings up of k-points. □

We return now to the proof of (7.12)(i). Let k be the local field, $\mathfrak{o} \subset k$ the ring of integers, and let

$$\tilde{X} \to \operatorname{Spec} \mathfrak{o}$$

be smooth and projective with special fibre X_0 and generic fibre X. Let k' be an unramified extension with residue field $\mathfrak{f}'$ splitting X_0. The units $\mathfrak{o}'^*$ in the ring of integers $\mathfrak{o}' \subset k'$ are cohomologically trivial [13] so $N \otimes \mathfrak{o}'^*$ is cohomologically trivial (Serre [13], theorem 9, p. 152). (Here $N = \operatorname{Pic}(X_{k'}) = \operatorname{Pic}(X_{\bar{k}})$.) The sequence

$$0 \to N \otimes \mathfrak{o}'^* \to N \otimes k'^* \to N \to 0$$

yields $H^1(k'/k, N \otimes k') \cong H^1(k'/k, N)$.

Because $N \cong \operatorname{Pic}(X_{0,\mathfrak{f}'})$, there is an exact sequence of $\operatorname{Gal}(k'/k)$ ($= \operatorname{Gal}(\mathfrak{f}'/\mathfrak{f})$) modules ($F'_0 = \mathfrak{f}'(X_0)$)

$$0 \to F_0'^* / \mathfrak{f}'^* \to \operatorname{Div}(X_{0,\mathfrak{f}'}) \to N \to 0.$$

Since $\mathrm{Div}(X_{0,\mathfrak{f}'})$ is a permutation module, we obtain a diagram with exact rows $(F' = k'(X))$

$$
\begin{array}{ccccc}
A_0(X) & \xrightarrow{\ \Phi\ } & H^1(k'/k, N\otimes k') & \xrightarrow{\ \theta(7.2)\ } & H^2(k'/k, K_2(F')/K_2(k')) \\
& & \| & & \downarrow{\scriptstyle\text{tame}} \\
0 & \longrightarrow & H^1(k'/k, N) & \xrightarrow{\ \partial\ } & H^2(k'/k, F_0'^{*}/\mathfrak{f}'^{*}).
\end{array}
$$

(7.12.2)

Here tame: $K_2(F') \to F_0'^{*}$ is the tame symbol associated to the divisor $X_{0,\mathfrak{f}'}$ on $\tilde{X} \times_{\mathrm{Spec}\,\mathfrak{o}} \mathrm{Spec}\,\mathfrak{o}'$. The fact that $\Phi = 0$ now follows from

Lemma (7.12.3) *The square in (7.12.2) is commutative up to sign.*

Proof A point $x' \in X_{k'}^1$ can be viewed as the generic point of a divisor $\{x'\}$ on $\tilde{X} \times \mathrm{Spec}\,\mathfrak{o}'$ flat over $\mathrm{Spec}\,\mathfrak{o}'$ with function field $k'(x')$. The intersection $\{x'\} \cap X_{0,\mathfrak{f}'}$ is a divisor on $\{x'\}$, so we obtain a valuation map

$$\alpha_{x'} : k'(x')^{*} \to \mathrm{Div}(X_{0,\mathfrak{f}'}).$$

Adding up over various x', we obtain a diagram

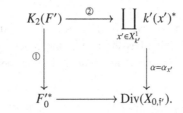

The maps labeled ① and ② are tame symbols and the diagram commutes up to a sign. Since $Z \subset \coprod k'(x')^{*}$ (cf. 7.10), we obtain a commutative diagram of Galois modules:

$$
\begin{array}{ccccccccc}
0 & \longrightarrow & K_2(F')/K_2(k') & \longrightarrow & Z & \longrightarrow & N \otimes k'^{*} & \longrightarrow & 0 \\
& & \downarrow{\scriptstyle①} & & \downarrow{\scriptstyle\alpha} & & \downarrow{\scriptstyle 1\otimes\text{valuation}} & & \\
0 & \longrightarrow & F_0'^{*}/\mathfrak{f}'^{*} & \longrightarrow & \mathrm{Div}(X_{0,\mathfrak{f}'}) & \longrightarrow & N & \longrightarrow & 0.
\end{array}
$$

The assertion regarding (7.12.2) is now straightforward. □

This completes the proof of (7.12). Returning to the exact sequence (7.2), we see that finiteness for $A_0(X)$ for X/k local or global will follow whenever we can prove $H^1(\bar{k}/k, K_2(\bar{F})/K_2(\bar{k}))$ is finite, for $\bar{F} = \bar{k}(X)$ and X a rational surface over k.

Definition (7.13) *X* is a *conic bundle* surface if there exists a rational pencil of genus-zero curves on *X*.

Examples (7.13.1) (i) The Châtelet surfaces discussed in (7.6) are conic bundle surfaces. More generally any surface

$$t_1^2 - f(x) t_2^2 = g(x)$$

is a conic bundle surface.

(ii) A cubic surface (smooth cubic hypersurface in $\mathbf{P}^3$) which contains a line ℓ defined over *k* is a conic bundle surface. Indeed, the residual intersection of *X* with a pencil of planes through ℓ is a pencil of conics.

Before stating our main result, we recall that the *Milnor ring* [8] $k_*(k)$ of a field *k* is the graded $\mathbf{Z}/2\mathbf{Z}$-algebra with generators elements $\ell(a) \in k_1(k)$, for $a \in k^*$, and relations $2\ell(a) = 0$, $\ell(ab) = \ell(a) + \ell(b)$, and $\ell(a)\ell(1 - a) = 0$, for $a \neq 0, 1$.

Theorem (7.14) *Let X be a conic bundle surface over a field k which is either a local field, a global field, or a C_i-field [13] for $i \leq 3$. Then the group $H^1(\bar{k}/k, K_2(\bar{F})/K_2(\bar{k}))$ is a subquotient of $k_3(k)$.*[1]

As a consequence of results in Milnor [8] one has

$$
k_3(k) = \begin{cases}
0 & k\ C_2\text{-field}, \\
0 & k \text{ local} \neq \mathbf{R}, \\
\mathbf{Z}/2\mathbf{Z} & k = \mathbf{R}, \\
\oplus_{v \text{ real place}} \mathbf{Z}/2\mathbf{Z} & k \text{ global}.
\end{cases}
$$

Combining (7.1), (7.13) and the above results of Milnor, one finds

Corollary (7.14.1) *If X is a conic bundle surface over a local or global field, then $A_0(X)$ is finite.*

Corollary (7.14.2) *If X is a cubic surface over a local or global field, then $A_0(X) \otimes_{\mathbf{Z}} \mathbf{Z}[\frac{1}{3}]$ is finite. In particular if X is defined over $\mathbf{R}$, then $A_0(X)$ is finite.*

Proof of (7.14.2) Let k'/k be a finite Galois extension splitting *X*. The existence of a transfer $A_0(X_{k'}) \to A_0(X_k)$ implies that $[k' : k]A_0(X_k) = (0)$. It suffices therefore to show the ℓ-power torsion subgroup $A_0(X)(\ell)$ is finite for any prime $\ell \neq 3$. Let $k_\ell \subset k'$ be the fixed field of the Sylow ℓ-subgroup of $\mathrm{Gal}(k'/k)$. Then ℓ does not divide $[k_\ell : k]$ so a transfer argument shows

---

[1] Combining their own ideas with arguments given in the text, Colliot-Thélène and Sansuc have shown $H^1(\bar{k}/k, K_2(\bar{F})/K_2(\bar{k})) \subset K_3(k)$ for conic bundle surfaces *X*, and this cohomology group is zero for *k* local or global unless $\chi(\mathbf{R}) = \phi$ for all real places of *k*. The group is non-zero for $k = \mathbf{R}$, $\chi = \mathbf{P}^1 \times C$, where *C* is the non-trivial conic over $\mathbf{R}$.

that $A_0(X)(\ell) \hookrightarrow A_0(X_{k_\ell})(\ell)$. We may assume, therefore, that $\mathrm{Gal}(k'/k)$ is an ℓ-group.

Recall $X_{\bar{k}}$ has 27 lines on it, all of which we may assume (taking k' large to begin with) are defined over k'. The ℓ-group $\mathrm{Gal}(k'/k)$ will necessarily leave a line fixed for $\ell \neq 3$, so X is a conic bundle surface (7.13.1)(ii).   □

Proof of (7.14)   Fix a conic bundle structure on X, that is a rational map $\pi: X \to \mathbf{P}^1_k$ with fibre of genus 0. Blowing up on X, we may assume π is everywhere regular. Let $K = k(\mathbf{P}^1)$ and $\bar{K} = \bar{k}(\mathbf{P}^1)$. Since $\bar{K}$ is a C_1-field [13] the generic fibre of π pulled back to $\bar{k}$, $X_{\bar{K}}$, is isomorphic to $\mathbf{P}^1_{\bar{K}}$. Writing $\bar{F} = \bar{k}(X) = \bar{K}(X_{\bar{K}})$ we have

$$(7.14.3) \qquad 0 \to K_2(\bar{K}) \to K_2(\bar{F}) \to \coprod_{\bar{x} \in X_{\bar{K}}} \bar{K}(\bar{x})^* \xrightarrow{N} \bar{K}^* \to 0,$$

where N is the norm map [7]. Similarly

$$(7.14.4) \qquad K_2(K) \cong K_2(\bar{k}) \oplus \coprod_{\bar{y} \in \mathbf{P}^1_k}^0 \bar{k}^*,$$

where as usual $\coprod^0 \bar{k}^* = \mathrm{Ker}(\coprod \bar{k}^* \to \bar{k}^*)$.

Let $\mathfrak{n} \subset K^*$ be the image of the norm map from

$$\left(\coprod_{\bar{x} \in X_{\bar{K}}} \bar{K}(\bar{x})^* \right)^{\mathrm{Gal}(\bar{k}/k)} = \coprod_{x \in X_k} K(x)^*.$$

From (7.14.3) and (7.14.4) we get an exact sequence

$$(7.14.5) \qquad 0 \to H^1(\bar{k}/k, K_2(\bar{F})/K_2(\bar{k})) \to K^*/\mathfrak{n} \xrightarrow{\psi} \coprod_{y \in \mathbf{P}^1_k} \mathrm{Br}(k(y)).$$

Fix $a, b \in K^*$ such that X_K is isomorphic to the conic curve

$$X_K : T_0^2 - aT_1^2 - bT_2^2 = 0.$$

We will work with $\ell(a) \cdot \ell(b) \in k_2(K)$, as well as with the quaternion algebra A over K defined by $T^2 = a$, $U^2 = b$, and $TU = -UT$.   □

Lemma (7.14.6)   *Let* $\mathrm{Nrd}: A^* \to K^*$ *be the reduced norm. Then* $\mathrm{Nrd}(A^*) \subset \mathfrak{n}$.

Proof   An extension K'/K splits A (i.e. $A \otimes_K K' \cong \mathrm{End}(K' \oplus K')$) if and only if X_K has a K'-point. If $\alpha \in A, \alpha \notin K$, then $K(\alpha)$ splits A, so X_K has a $K(\alpha)$-point. Since the reduced norm on A coincides on $K(\alpha) \subset A$ with the field norm, we get $\mathrm{Nrd}(\alpha) \in \mathfrak{n}$. Finally, if $\alpha \in K$, $\mathrm{Nrd}(\alpha) = \alpha^2 \in \mathfrak{n}$, since X_K obviously contains points of degree 2 over K.   □

We now can rewrite (7.14.5)

$$0 \longrightarrow H^1(\bar{k}/k, K_2(\bar{F})/K_2(\bar{k})) \longrightarrow K^*/\mathfrak{n} \overset{\psi}{\longrightarrow} \coprod_{y \in \mathbf{P}^1_k} Br(k(y))$$

(7.14.7)

$$K^*/Nrd(A^*).$$

and it will suffice to show $\operatorname{Ker} \psi' \hookrightarrow k_3(k)$.

Lemma (7.14.8) *The map* $\ell(a)\cdot\ell(b)\colon K^* \to k_3(K)$, *where* $c \mapsto \ell(c)\cdot\ell(a)\cdot\ell(b)$, *contains* $Nrd(A^*)$ *in its kernel.*

Proof Let $\beta = Nrd(\alpha) \in Nrd(A^*)$. We may assume $\alpha \notin K^*$, so $K(\alpha)$ splits A. This implies $\ell(a) \cdot \ell(b) \to 0$ in $k_2(K(\alpha))$ (Milnor [9], p. 152). The projection formula [7] implies

$$\ell(Nrd(\alpha)) \cdot \ell(b)\ell(a) = N_{K(\alpha)/K}(\ell(\alpha)) \cdot \ell(b) \cdot \ell(a)$$
$$= N_{K(\alpha)/K}(\ell(\alpha) \cdot \ell(b) \cdot \ell(a)) = 0. \qquad \square$$

Lemma (7.14.9) *The diagram*

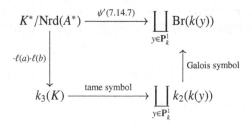

commutes.

Lemma (7.14.10) *The maps in the above square satisfy:*

(i) *"Galois symbol" is injective,*

(ii) *"tame symbol" has kernel* $k_3(k) \subset k_3(K)$,

(iii) *"*$\cdot \ell(a) \cdot \ell(b)$*" is injective.*

Note that these two lemmas suffice to complete the proof of (7.14). Note also their geometric content. Let $\sum n_i(y_i)$ be a zero-cycle of degree zero on $\mathbf{P}^1_k$. Assume the fibres of $\pi\colon X \to \mathbf{P}^1$ over the y_i are all smooth. A necessary condition for there to exist a cycle Z on X such that $\pi_* z = \sum n_i(y_i)$ is that n_i should be even whenever $\pi^{-1}(y_i)$ is non-split (i.e. $\not\cong \mathbf{P}^1_{k(y_i)}$). If $(f) = \sum n_i(y_i)$,

(7.14.9) implies

$$\psi'(f) = \sum n_i [\pi^{-1}(y_i)], \quad \text{where } [\pi^{-1}(y_i)] = \text{class of}$$
$$\text{Severi–Brauer variety in } \mathrm{Br}(k(y_i)).$$

Thus, in the local or global case when $k_3(k)$ is finite, we have divisors of degree zero and even order at non-split places modulo norms of zero-cycles rationally equivalent to on X = finite group. Finally the reader may wish to compare this chain of ideas with the Eichler norm theorem describing the image of the norm mapping for a divisor algebra over a local or global field.

Proof of (7.14.10)   Injectivity of the Galois symbol follows from Lam [10], given our assumptions about k, and the fact that Ker(tame) = $k_3(k)$ is proven in [8]. To show injectivity of multiplication by $\ell(a) \cdot \ell(b)$, we consider the map defined in Milnor [8]

$$k_3(K) \to I^3/I^4,$$
$$\ell(a)\ell(b)\ell(c) \to (\langle c \rangle + \langle -1 \rangle)(\langle b \rangle + \langle -1 \rangle)(\langle a \rangle + \langle -1 \rangle),$$

where I is the augmentation ideal in the *Witt ring* of K. The above quadratic form is a *Pfister form* in the terminology of [10]. In particular, this form lies in I^4 if and only if it is *hyperbolic* ([10] cor. 3.4, p. 290; thm. X.6.17, p. 367, in the 2005 book). Thus $\ell(a)\ell(b)\ell(c) = 0$ implies there exist $x_1, \ldots, x_8 \in K$ not all 0 such that

$$abc\, x_1^2 - ab\, x_2^2 - ac\, x_3^2 - bc\, x_4^2 + a\, x_5^2 + b\, x_6^2 + c\, x_7^2 - x_8^2 = 0.$$

Formally, then, we may factor and write

$$c = \frac{abx_2^2 - ax_5^2 - bx_6^2 + x_8^2}{abx_1^2 - ax_3^2 - bx_4^2 + x_7^2}.$$

Notice both numerator and denominator are norms from A^*. We may assume $A \not\cong \mathrm{End}(K \oplus K)$, else $X_K \cong \mathbf{P}_K^1$, X is k-rational, and the whole discussion is silly. Thus the denominator in the above expression vanishes if and only if $x_1 = x_3 = x_4 = x_7 = 0$. But vanishing of the denominator implies vanishing of the numerator and hence all the x_i, a contradiction. Hence neither numerator nor denominator vanishes and we have written c as a norm from A^*. $\square$

Proof of (7.14.9)

Lemma (7.14.11)   *Let X_K be a conic curve over* Spec K *with $\bar{K}/K$ Galois splitting X_K. Let $[X_K] \in H^2(\bar{K}/K, \bar{K}^*)$ be the class of X_K as a Severi–Brauer* *variety, and assume* $[X_K] \neq 0$. *Let* $\bar{F} = $ *quotient field of* $X_{\bar{K}}$ *and consider the exact sequence of* $\mathrm{Gal}(\bar{K}/K)$*-modules*

$$0 \to \bar{K}^* \to \bar{F}^* \to \coprod_{x \in X_{\bar{K}}} \mathbf{Z} \to \mathbf{Z} \to 0.$$

Then $[X_K] = \partial_2 \partial_1(1)$, *where* ∂_i *are the boundary maps associated to this exact sequence.*

Proof

$$\left(\coprod_{x \in X_{\bar{K}}} \mathbf{Z} \right)^{\mathrm{Gal}(\bar{K}/K)} = \coprod_{x \in X_K} \mathbf{Z}.$$

Since any $x \in X_K$ has even degree over $\mathrm{Spec}\,K$ (this follows from a norm argument using the fact that X_K splits over $K(x)$), we get an exact sequence

$$0 \to \mathbf{Z}/2\mathbf{Z} \to H^2(\bar{K}/K, \bar{K}^*) \to H^2(\bar{F}/F, \bar{F}^*),$$
$$1 \mapsto \partial_2 \partial_2(1).$$

Since $[X_K] \to 0$ in $H^2(\bar{F}/F, \bar{F}^*)$, we are done. □

Tensoring the sequence in (7.14.11) with K^* and using the symbol map, we obtain a commutative diagram

$$0 \to \bar{K}^* \otimes_{\mathbf{Z}} K^* \to \bar{F}^* \otimes K^* \to \coprod_{\bar{x} \in X^1_{\bar{K}}} K^* \to K^* \to 0$$
$$0 \to K_2(\bar{K}) \to K_2(\bar{F}) \to \coprod_{\bar{x}} \bar{K}(\bar{x})^* \to \bar{K}^* \to 0.$$

Writing ∂'_i for the boundary maps on cohomology associated to the bottom row, we find a commutative triangle

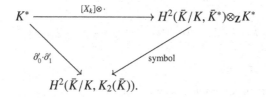

$$K^* \xrightarrow{[X_k] \otimes \cdot} H^2(\bar{K}/K, \bar{K}^*) \otimes_{\mathbf{Z}} K^*$$
$$\partial'_0 \cdot \partial'_1 \searrow \qquad \swarrow \text{symbol}$$
$$H^2(\bar{K}/K, K_2(\bar{K})).$$

Now take $K = k(\mathbf{P}^1)$, $\bar{K} = \bar{k}(\mathbf{P}^1)$ and compose with the tame symbol to get a

commutative square

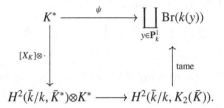

The proof of (7.14.9) now follows from

Lemma (7.14.12) *Let k be a field of characteristic $\neq 2$, $\bar{k} =$ separable closure of k. Let K be an extension field of transcendence degree 1 over k, and write $\bar{K} = K\bar{k}$. Let $a, b \in K^*$ and write $\ell(a)\ell(b) \in k_2(K)$, $(a, b) \in {}_2\mathrm{Br}(K)$. Then*

$$\mathrm{Br}(K) \cong H^2(\bar{k}/k, \bar{K}^*).$$

Moreover, if y is a place of K over k with residue field $k(y)$, the diagram

(7.14.13)

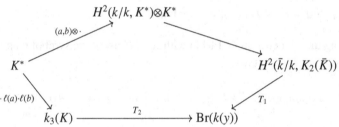

(where T_1 and T_2 are tame symbols) commutes.

Proof Replacing k by its perfect closure, we may assume $k(y)$ separable over k. Next replacing K by its completion at y we may suppose $k(y) \subset K$ and $\bar{K} = \coprod \bar{K}_i$, one copy for each place lying over y. Now replacing the Galois group by the decomposition group for one of the $\bar{K}_i$ we may assume $k = k(y)$, $K = k((\pi))$.

We now have split exact sequences

$$0 \longrightarrow \mathfrak{o}^* \longrightarrow K^* \longrightarrow \mathbf{Z} \longrightarrow 0, \quad 0 \longrightarrow \bar{\mathfrak{o}}^* \longrightarrow \bar{K}^* \longrightarrow \mathbf{Z} \longrightarrow 0.$$

$$\pi \longleftarrow\!\shortmid 1$$

Note that $1 + \pi\bar{\mathfrak{o}}$ is cohomologically trivial, so

$$H^2(\bar{k}/k, \bar{K}^*) \cong H^2(\bar{k}/k, \bar{k}^*) \oplus H^2(\bar{k}/k, \mathbf{Z}). \qquad \square$$

Case 1 For $a, b \in \mathfrak{o}$, let $a_0, b_0 \in k^*$ denote the mod π reductions of a, b. In this case $(a, b) = (a_0, b_0) \in H^2(\bar{k}/k, \bar{k}^*)$. It is easy enough to see that going either way around (7.14.13), $f \in K^*$ gets taken to $\text{ord}(f) \cdot (a_0, b_0) \in \text{Br}(k)$.

Using linearity, it remains only to consider

Case 2 Suppose $b = \pi$, $a \in \mathfrak{o}^*$. In this case let $G = \text{Gal}(\bar{k}/k)$ and let $\rho :$ $G \times G \to \mathbf{Z}$ be a 2-cocycle representing the image of a under the composition

$$(7.14.14) \qquad K^* \to H^1(\bar{K}/K, \mu_2) \xrightarrow{\delta} H^2(\bar{K}/K, \mathbf{Z}).$$
$$a \mapsto \chi_a$$

Here δ is the coboundary from the exact sequence

$$0 \to \mathbf{Z} \to \mathbf{Z} \to \mathbf{Z}/2\mathbf{Z} \to 0.$$

Then (a, b) is represented by the cocycle

$$G \times G \to \bar{K}^*, \qquad (g_1, g_2) \mapsto \pi^{\rho(g_1, g_2)}.$$

Indeed, one knows (cf. Serre [13], p. 214) that

$$(a, \pi) = (\delta \chi_a) \cdot \pi^{\rho(\cdot, \cdot)}.$$

With reference to (7.14.13), we must show

$$T_1\{\pi^{\rho(g_1, g_2)}, f\} = T_2(\ell(a)\ell(\pi)\ell(f)).$$

If $F = -\pi$ this is clear as both sides are trivial. If, on the other hand, f is a unit with residue class f_0, we reduce to showing that $\ell(a_0)\ell(f_0) \in k_2(k)$ maps to the element in $\text{Br}(k)$ represented by the cocycle $f_0^{\rho(g_1, g_2)}$. This follows as above with K in (7.14.14) replaced by k and $\bar{K}$ replaced by $\bar{k}$. This completes the proof of (7.14.12) and (7.14.9).

$$\square$$

References for Lecture 7

There is a vast literature about points on rational surfaces. A brief list of works related in one way or another to the point of view of this lecture follows:

[1] F. Châtelet, Points rationnels sur certaines courbes et surfaces cubiques, *Enseignement Math. (2)*, **5** (1959), 153–170 (1960).

[2] J.-L. Colliot-Thélène and J.-J. Sansuc, Series of notes on rational varieties and groups of multiplicative type, *C. R. Acad. Sci. Paris Ser. A-B*, **282** (1976), A1113–A1116; **284** (1977), A967–A970; **284** (1977), A1215–A1218; **287** (1978), A449–A452.

[3] J.-L. Colliot-Thélène and J.-J. Sansuc, La *R*-équivalence sur les tores, *Ann. Sci. École Norm. Sup. (4)*, **10** (1977), 175–229.

[4] J.-L. Colliot-Thélène and D. Coray, L'équivalence rationnelle sur les points fermés des surfaces rationnelles fibrées en coniques, *Compositio Math.*, bf 39 (1979), 301–332.

[5] Yu. Manin, *Cubic Forms*, North Holland, Amsterdam (1974). [Second edition, 1986.]

[6] Yu. Manin, Le groupe de Brauer–Grothendieck en géométrie diophantienne, pp. 401–411 in *Actes du Congrès International Mathématiciens (Nice, 1970)*, vol. 1, Gauthier-Villars, Paris (1971).

For a more extensive list, the reader can see the bibliography in [5].

The following are referred to in the text for the arithmetic of symbols in K-theory, and for quadratic forms:

[7] H. Bass and J. Tate, The Milnor ring of a global field, pp. 349–446 in *Algebraic K-Theory II*, Lecture Notes in Math., no. 342, Springer, Berlin (1973).

[8] J. Milnor, Algebraic K-theory and quadratic forms, *Invent. Math.*, **9** (1970), 318–344.

[9] J. Milnor, *Introduction to Algebraic K-Theory*, Annals of Mathematics Studies, vol. 72, Princeton University Press, Princeton, N.J. (1971).

[10] T. Y. Lam, *The Algebraic Theory of Quadratic Forms*, W. A. Benjamin, Reading, Mass. (1973). [Revised second printing, 1980. See also *Introduction to Quadratic Forms over Fields*, American Mathematical Society, Providence, R.I., 2005.]

[11] A. Grothendieck, Le groupe de Brauer I, II, III, pp. 46–188 in *Dix exposés sur la cohomologie des schémas*, North Holland, Amsterdam (1968).

[12] T. Nakayama, Cohomology of class field theory and tensor product modules I, *Ann. of Math. (2)*, **65** (1957), 255–267.

[13] J. P. Serre, *Corps Locaux*, second edition, Hermann, Paris (1968). [Translation: *Local Fields*, Springer, New York, 1979.]

[14] H. Gillet, Applications of algebraic K-theory to intersection theory, Thesis, Harvard (1978).

[15] C. Sherman, K-cohomology of regular schemes, *Comm. Algebra*, **7** (1979), 999–1027.

8

Relative cycles and zeta functions

We return now to the relative cycle maps described in Lecture 3. Changing notation, we write

$$j : \left\{ \begin{array}{l} \text{relative algebraic} \\ \text{1-cycles on } C \times \mathbf{P}^1 \times \mathbf{P}^1 \end{array} \right\} \to H^1(C, \mathbf{C}^*),$$

and

$$\log|j| : \left\{ \begin{array}{l} \text{algebraic 1-cycles on} \\ C \times \mathbf{P}^1 \times \mathbf{P}^1 \text{ meeting} \\ C \times \# \text{ properly} \end{array} \right\} \to H^1(C, \mathbf{R})$$

for the cycle class maps. (Here C is a smooth complete curve over a field $k \hookrightarrow \mathbf{C}$.) When $C = E$ is an elliptic curve we will show explicitly how to write down some interesting relative cycles γ and we will compute $\log|j|(\gamma)$. In Lecture 9 we consider the case when $E/\mathbf{Q}$ has complex multiplication by the full ring of integers in an imaginary quadratic number field. Identifying $H^1(E, \mathbf{R}) \cong \mathbf{C}$, we show there is a natural relative cycle γ defined over $\mathbf{Q}$ such that

$$\rho \cdot \log|j|(\gamma) = L(2, \chi^{\text{Gröss}}),$$

where $L(s, \chi^{\text{Gröss}})$ is the Artin–Hasse zeta function of E (given in this case by a Hecke L-series, whence the notation) and ρ is a constant built from a Gauss sum, the conductor of E, and the imaginary part of τ. Since $L(s, \chi)$ has an Euler product converging for Re$s > 3/2$, it will follow that $\log|j|(\gamma) \neq 0$.

These computations mimic those in an unpublished preprint [2], where I define a map from K-theory

$$\Gamma(E, \mathcal{K}_2) \to H^1(E(\mathbf{C}), \mathbf{R}) \cong \mathbf{C}.$$

It seems likely that $\Gamma(E, \mathcal{K}_2)$ can be identified with a suitable Chow group for

$E \times \mathbf{P}^1 \times \mathbf{P}^1$ relative to $E \times \#$ (cycles being defined over $\mathbf{Q}$) and that the rank of this group is the order of vanishing of $L(s, \chi)$ at $s = 0$. Indeed, one may hope for a similar interplay between relative cycles, K-groups, and zeta functions quite generally. I hope this exciting prospect will sustain the reader through the complicated calculations which follow.

Recall we have an exact sequence

$$0 \longrightarrow \Gamma(C, \mathcal{K}_2) \longrightarrow K_2(k(C)) \overset{T}{\longrightarrow} \coprod_{x \in C} k(x)^*.$$

Our first objective will be to associate to an element in $\Gamma(C, \mathcal{K}_2)$ a relative algebraic 1-cycle on $C \times \mathbf{P}^1 \times \mathbf{P}^1$, relative to $C \times \#$. This cycle will only be well defined modulo some equivalence which we shall not describe explicitly.

Consider then an element $\prod\{f_i, g_i\} \in \Gamma(C, \mathcal{K}_2) \subset K_2(k(C))$. To each pair f, g of rational functions on C we associate the double graph

$$\gamma_{f,g} = \{(x, f(x), g(x))\} \subset C \times \mathbf{P}^1 \times \mathbf{P}^1.$$

Let $\gamma_1 = \sum \gamma_{f_i, g_i}$. The name of the game will be to add or subtract "trivial" curves to γ_1 in such a way that the resulting sum is a relative algebraic 1-cycle. A curve $\tau \subset C \times \mathbf{P}^1 \times \mathbf{P}^1$ is "trivial" for our purposes if either $\tau \subset \{p\} \times \mathbf{P}^1 \times \mathbf{P}^1$ for some $p \in C$, or $\tau \subset (C \times \mathbf{P}^1 \times \{1\}) \cup (C \times \{1\} \times \mathbf{P}^1)$. One sees from Lecture 3 that $\log |j|$ (trivial curve) $= 0$.

Step 1 (Neutralizing the vertices)   It may happen that Supp γ_1 contains a point (p, i, j), $i, j = 0, \infty$. Suppose for example one of the components of γ_1 looks near $(p, 0, 0)$ like

$$(8.1.1) \qquad \{(x, ax^r + \text{hot}, bx^s + \text{hot})\}, \quad \text{hot} = \text{higher order terms},$$

where x is a local coordinate near p, $x(p) = 0$. A calculation like (3.2) shows that this component can be made to look like a relative cycle near $(p, 0, 0)$ by subtracting the "trivial" curve

$$(8.1.2) \qquad \tau = \left\{\left(p, at^r, b\left(\frac{t}{1-t}\right)^s\right)\right\}, \quad t = \text{standard parameter on } \mathbf{P}^1.$$

Note that in addition to the intersection at $(p, 0, 0)$, τ meets $C \times \#$ with multiplicity r at $(p, \infty, (-1)^s b)$ and with multiplicity s at (p, a, ∞). We iterate the above procedure *mutatis mutandis* for every vertex point (p, i, j) on γ_1.

Step 2 (Neutralizing other boundary points)   Let $\gamma_2 = \gamma_1 - \sum \tau$ be the cycle with vertex intersections neutralized. Recall we have on $K_2(k(C))$ the tame symbol at p

$$T_p \colon K_2(k(C)) \to k(p)^*,$$

$$T_p\{f, g\} = (-1)^{\operatorname{ord}_p f \cdot \operatorname{ord}_p g} \frac{f^{\operatorname{ord}_p g}}{g^{\operatorname{ord}_p f}} (p).$$

The tame symbol can be interpreted geometrically as follows: let

$$W = C \times \mathbf{G}_m \times \{0\} - C \times \mathbf{G}_m \times \{\infty\} - C \times \{0\} \times \mathbf{G}_m + C \times \{\infty\} \times \mathbf{G}_m$$

(note we have removed the vertices). For $w \in \operatorname{Supp} W$, let $\rho(w) \in \mathbf{G}_m$ be either the second or third coordinate, whichever $\neq 0, \infty$. For $x \in C \times \mathbf{P}^1 \times \mathbf{P}^1$ let $\nu(\gamma_2, W; x)$ be the multiplicity of intersection of γ_2 and W at x.

Lemma (8.1.3) *With notation as above,*

$$1 = T_p\left(\prod_i \{f_i, g_i\}\right) = \prod_{w \in (\{p\} \times \mathbf{P}^1 \times \mathbf{P}^1) \cap \operatorname{Supp} W} \rho(w)^{\nu(\gamma_2, W; w)}.$$

Proof A cycle like (8.1.1) contributes $(-1)^{rs} a^s b^{-r}$ to the tame symbol. This is precisely the contribution of the cycle $-\tau$ to the right-hand side, where τ is as in (8.1.2). The bookkeeping for non-vertex points is easier, and is left for the reader. □

Given $a, b \in \mathbf{G}_m$ and $p \in C$, we consider the "trivial" cycles

$$A_{a,b,p} = \left\{\left(p, t, \frac{(t-a)(t-b)}{(t-1)(t-ab)}\right) \,\middle|\, t \in \mathbf{P}^1\right\},$$

$$B_{a,b,p} = \left\{\left(p, t, \frac{t-a}{t-b}\right) \,\middle|\, t \in \mathbf{P}^1\right\}.$$

We have for $\bar{W} = $ closure of W, viewed as a cycle,

$$A_{a,b,p} \cdot \bar{W} = (p, a, 0) + (p, b, 0) - (p, 0, 1) - (p, ab, \infty)$$
$$- (p, 1, \infty) + (p, \infty, 1),$$

$$B_{a,b,p} \cdot \bar{W} = (p, a, 0) - (p, b, \infty) - \left(p, 0, \frac{a}{b}\right) + (p, \infty, 1).$$

Starting at these formulae and using (8.1.3), it is straightforward to construct a cycle

$$\gamma_3 = \gamma_2 + \sum n_{ij} A_{a_{ij}, b_{ij}, p_i} + \sum m_{rs} B_{a_{rs}, b_{rs}, p_r}$$

such that $\gamma_3 \cdot (C \times \#)$ has support in

$$(C \times \{1\} \times \{0, \infty\}) \cup (C \times \{0, \infty\} \times \{1\}).$$

Step 3 (Finishing the job)   We may now add or subtract cycles of the form $(p) \times (1) \times \mathbf{P}^1$ and $(p) \times \mathbf{P}^1 \times (1)$ to γ_3 getting a γ_4 such that $\gamma_4 \cdot C \times (i) \times \mathbf{P}^1$ and $\gamma_4 \cdot C \times \mathbf{P}^1 \times (i)$ are effective, $i = 0, \infty$, but no cycle $\gamma_4 - (p) \times \mathbf{P}^1 \times (1)$ or $\gamma_4 - (p) \times (1) \times \mathbf{P}^1$ has this property. Write $\mathrm{pr}_1 : C \times \mathbf{P}^1 \times \mathbf{P}^1 \to C$. We have that

$$\mathrm{pr}_{1*}[\gamma_4 \cdot C \times ((0) - (\infty)) \times \mathbf{P}^1],$$
$$\mathrm{pr}_{1*}[\gamma_4 \cdot C \times \mathbf{P}^1 \times ((0) - (\infty))]$$

are zero-cycles rationally equivalent to 0 on C. There exist, therefore, functions h_1, h_2 on C such that

$$\gamma = \gamma_5 = \gamma_4 + \{(p, h_1(p), 1) \mid p \in C\} + \{(p, 1, h_2(p)) \mid p \in C\}$$

is a relative cycle.

We now want to construct some interesting elements in $\Gamma(C, \mathcal{K}_2)$ (and hence some interesting relative algebraic 1-cycles) *in the case $C = E = $ elliptic curve*. Suppose E is defined over a field k. Let f, g be functions on E, $N > 0$ an integer, and assume all zeros and poles of f and g are points of E of order N defined over k. Consider the diagram

(8.2)

$$
\begin{array}{ccccccc}
k(E)^* \otimes k^* & \xrightarrow{\mathrm{div} \otimes 1} & \coprod_{x \in E} k^* & \longrightarrow & \mathrm{Pic}(E) \otimes k^* & \longrightarrow & 0 \\
\Big\downarrow {\scriptstyle \mathrm{symbol}} & & \Big\downarrow & & & & \\
K_2(k(E)) & \xrightarrow{\mathrm{tame}} & \coprod_{x \in E} k(x)^*. & & & &
\end{array}
$$

Clearly $\mathrm{tame}\{f, g\}^N \in \coprod k^*$ and goes to 0 in $\mathrm{Pic}(E) \otimes k^*$. There thus exist functions $f_i \in k(E)^*$ and constants $c_i \in k^*$ such that

$$\{f, g\}^N \prod \{f_i, c_i\} \in \Gamma(E, K_2) .$$

Notation (8.2.1)   Fix an integer N and assume all points on E of order N are defined over k. If a is such a point, let f_a denote the function with zero of order N at a, and pole of order N at 0. Let ρ be the function with poles of order 1 at every non-zero point of order N, and a zero of order $N^2 - 1$ at the origin. Define $S_a \in \Gamma(E, K_2)$ by

$$S_a \equiv \{\rho, f_a\}^N \mod \mathrm{Image}(k(E)^* \otimes k^* \to K_2(k(E))) .$$

S_a is actually well defined modulo torsion and symbols with both entries constant, as one sees by looking at the kernel of $\mathrm{div} \otimes 1$ in (8.2).

Our objective is now to compute $\log |j(S_a)|$. The key step in this program will be to compute $\log |j(\gamma_{f,g})|$ when $\gamma_{f,g} = \{(x, f(x), g(x)) \mid x \in E\}$ for any two functions f, g on E. We assume for simplicity that the divisors of f and g have disjoint support. (The general case follows from this one by a limit argument.) Slit E along non-intersecting simple closed paths between zeros and poles of f, removing also ε-disks about these points:

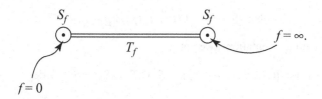

We do likewise for the zeros and poles of g, taking care that the slits for f and g do not intersect. (This is possible because removing a simple arc will not disconnect the surface.) The resulting slit and punctured surface will be denoted E'. The handles of the barbell-shaped affairs are called T_f, T_g or $T_{f,\varepsilon}, T_{g,\varepsilon}$. The ε-circles are written S_f, S_g or $S_{f,\varepsilon}, S_{g,\varepsilon}$.

It will be convenient to use the Jacobi parameterization $\pi \colon \mathbf{C}^* \to E \cong \mathbf{C}^*/q^{\mathbf{Z}}$, $q = e^{2\pi i \tau}$. We let t be the standard parameter on $\mathbf{C}^*$ and write $t = e^{2\pi i z}$, $z = x + iy$. Thus $y = \frac{-1}{2\pi} \log |t|$ is a function on $\mathbf{C}^*$, but x is not.

We normalize the Weierstrass σ-function on E as in Lang [4]

$$\phi(z) = e^{-\frac{1}{2}\eta z^2 + \pi i z}, \quad \sigma(z) = \frac{1}{2\pi i}(t-1)\prod_{n=1}^{\infty} \frac{(1 - q^n t)(1 - q^n t^{-1})}{(1 - q^n)^2}.$$

The given functions f, g on E can be factored in terms of the ϕ-function

$$f(z) = c_f \prod \phi(z - a_j)^{m_j}, \quad g(z) = c_g \prod \phi(z - b_k)^{n_k}, \quad c_f, c_g \in \mathbf{C}^*,$$

$$\sum m_j = \sum n_k = \sum m_j a_j = \sum n_k b_k = 0.$$

For any fixed n we have

$$\prod_k e^{2\pi i n_k(z - a_k - n\tau)} = 1,$$

so if we write

$$\tilde{f} = \prod_j (1 - \alpha_j t)^{n_j}, \quad \alpha_j = e^{-2\pi i a_j},$$

we get

$$f(t) = c_f \prod_{n \in \mathbf{Z}} \tilde{f}(tq^n).$$

Finally, let $\mathbf{C}^{*\prime}$ denote $\mathbf{C}^*$ slit and punctured so as to lie over E', $\pi \colon \mathbf{C}^{*\prime} \to E'$. Functions like $\log f$ and $\log g$ (resp. $\log f$, $\log \tilde{f}$, $\log g$) can be defined on E' (resp. $\mathbf{C}^{*\prime}$). Also a family of paths $\overrightarrow{1f(p)}$ or $\overrightarrow{1g(p)}$ lying on $\mathbf{G}_m$ can be chosen continuously for $p \in E'$.

Define 3-chains

$$\Delta = \Delta_\varepsilon = \{(p, \overrightarrow{1f(p)}, g(p)) \mid p \in E'\},$$

$$\Phi = \Phi_\varepsilon = \{(p, \overrightarrow{1f(p)}, \overrightarrow{1g(p)}) \mid p \in T_f\}.$$

Bearing in mind orientations, we get

$$\partial\Delta = \{(p, \overrightarrow{1f(p)}, g(p)) \mid p \in S_g \cup T_f \cup S_f\} = \gamma'_{f,g} + \deg,$$

$$\partial\Phi = \{(p, \overrightarrow{1f(p)}, \overrightarrow{1g(p)}) \mid p \in \partial T_f\}$$
$$+ \{(p, \overrightarrow{1f(p)}, g(p)) \mid p \in T_f\} + \deg,$$

(where $\deg = $ cycle supported on $E \times \mathbf{G}_m \times \{1\} \cup E \times \{1\} \times \mathbf{G}_m$), whence $\partial(\Phi - \Delta) = \gamma'_{f,g} + $ chains supported on $(S_g \cup S_f) \times \mathbf{G}_m \times \mathbf{G}_m + $ degenerate chains.

Our duty becomes the calculation of

$$\lim_{\varepsilon \to 0} \frac{-1}{4\pi^2} \operatorname{Im} \int_{\Phi_\varepsilon - \Delta_\varepsilon} dx \wedge \frac{dx_1}{x_1} \wedge \frac{dx_2}{x_2}$$

and

$$\lim_{\varepsilon \to 0} \frac{-1}{4\pi^2} \operatorname{Im} \int_{\Phi_\varepsilon - \Delta_\varepsilon} dy \wedge \frac{dx_1}{x_1} \wedge \frac{dx_2}{x_2}.$$

Lemma (8.3)

$$\lim_{\varepsilon \to 0} \int_{\Delta_\varepsilon} dy \wedge \frac{dx_1}{x_1} \wedge \frac{dx_2}{x_2} = 2\pi i \left(y \log(1-t) \Big|_{\mathfrak{n}} - \int_{\mathfrak{n}} \log(1-t)\,dy \right).$$

(Here $\mathfrak{n}$ denotes the infinite "cycle" on $\mathbf{C}^*$

$$\mathfrak{n} = \sum_{\substack{n \in \mathbb{Z} \\ j,k}} m_j\, n_k(\alpha_k \beta_j^{-1} q^n), \qquad \alpha_k = e^{-2\pi i a_k}, \qquad \beta_j = e^{-2\pi i b_j},$$

and $\int_{\mathfrak{n}}$ denotes the integral along paths from points of negative multiplicity to points of positive multiplicity. A different choice of paths does not change the right-hand side.)

Proof of lemma   The first claim is that

$$\lim_{\varepsilon \to 0} \int_{E'_\varepsilon} dy \wedge \frac{dg}{g} = 0.$$

To see this, think of $g: E \to \mathbf{P}^1$. The above integral becomes

$$\lim_{\varepsilon \to 0} \left(-\frac{i}{2}\right) \int_{\mathbf{P}^1_\varepsilon} g_*(dz - d\bar{z}) \frac{du}{u},$$

where $\mathbf{P}^1_\varepsilon$ denotes $\mathbf{P}^1$ with ε-balls about 0 and ∞ removed, and u is the standard parameter on $\mathbf{P}^1$. Note that $g_*(dz)$ (resp. $g_*(d\bar{z})$) are global holomorphic (resp. anti-holomorphic) differentials on $\mathbf{P}^1$, and hence are zero, so the integral vanishes.

Now write

$$\int_{\Delta_\varepsilon} dy \wedge \frac{dx_1}{x_1} \wedge \frac{dx_2}{x_2} = -\int_{E'_\varepsilon} \log f \, dy \wedge \frac{dx_2}{x_2}.$$

Note that the differential can be written

$$\log f \, dy \wedge \frac{dx_2}{x_2} = c_f \, dy \wedge \frac{dx_2}{x_2} + \text{``trace''}_{C^*/E} \log \tilde{f} \, dy \wedge \frac{dx_2}{x_2},$$

where "trace" means the infinite sum over all deck transformations. The integral over $c_f \, dy \wedge \frac{dx_2}{x_2}$ vanishes by the above, so

$$\int_{\Delta_\varepsilon} dy \wedge \frac{dx_1}{x_1} \wedge \frac{dx_2}{x_2} = -\int_{C''} \log \tilde{f} \, dy \wedge \frac{dg}{g}$$
$$= -\int_{C''} d(y \log \tilde{f}) \wedge \frac{dg}{g} = \int_{T_f \cup T_g \cup S_f \cup S_g} \log g \, d(y \log \tilde{f}).$$

Winding around a zero of g changes $\log g$ by $2\pi i$, whence

$$\int_{T_g} \log g \, d(y \log \tilde{f}) = -2\pi i \int_{(g)} d(y \log \tilde{f}) = -2\pi i \, y \log \tilde{f} \Big|_{(g)},$$

where $\int_{(g)}$ means integrals taken over paths from poles to zeros of g. Similar arguments show

$$\lim_{\varepsilon \to 0} \int_{T_f} \log g \, d(y \log \tilde{f}) = -2\pi i \int_{(\tilde{f})} \log g \, dy.$$

Standard estimates imply

$$\lim_{\varepsilon \to 0} \int_{S_g} \log g \, d(y \log \tilde{f}) = 0,$$

and an easy argument (replacing y by $(y - y_0) + y_0$ where y_0 is the value of y at the center of a circle $S \subset S_f$) gives

$$\lim_{\varepsilon \to 0} \int_{S_f} \log g \, d(y \log \tilde{f}) = \lim_{\varepsilon \to 0} \int_{S_f} y \log g \, \frac{d\tilde{f}}{f} = 2\pi i \, y \log g \Big|_{(\tilde{f})}.$$

Combining these calculations

$$\lim_{\varepsilon \to 0} \int_{\Delta_\varepsilon} dy \wedge \frac{dx_1}{x_1} \wedge \frac{dx_2}{x_2} = 2\pi i \left(y \log g \Big|_{(\tilde{f})} - y \log f \Big|_{(g)} - \int_{(\tilde{f})} \log g \, dy \right).$$

We have $(g) = \sum_{n \in \mathbf{Z}, k} n_k(\beta_k^{-1} q^n)$ and $(\tilde{f}) = \sum_j m_j(\alpha_j^{-1})$, so

$$\lim_{\varepsilon \to 0} \int_{\Delta_\varepsilon} dy \wedge \frac{dx_1}{x_1} \wedge \frac{dx_2}{x_2}$$

$$= 2\pi i \left(\sum_{\substack{j,k \\ n \in \mathbf{Z}}} \frac{-m_j n_k}{2\pi} \log |\alpha_j^{-1}| \log (1 - \beta_k q^n \alpha_j^{-1}) \right.$$

$$\left. + \sum_{\substack{j,k \\ n \in \mathbf{Z}}} \frac{m_j n_k}{2\pi} \log |q^n \beta_k^{-1}| \log (1 - \alpha_j \beta_k^{-1} q^n) - \int_\eta \log(1 - t) \, dy \right).$$

Since for fixed n, k

$$\sum_j m_j \log \left(- \alpha_j \beta_k^{-1} q^n \right) = 0,$$

the above sum can be rewritten (replacing n by $-n$ in the second sum)

$$\lim_{\varepsilon \to 0} \int_{\Delta_\varepsilon} dy \wedge \frac{dx_1}{x_1} \wedge \frac{dx_2}{x_2} = +2\pi i \left(y \log(1 - t) \Big|_\eta - \int_\eta \log(1 - t) \, dy \right).$$

This completes the proof of Lemma (8.3).   □

Lemma (8.4)

$$\lim_{\varepsilon \to 0} \int_{\Delta_\varepsilon} dx \wedge \frac{dx_1}{x_1} \wedge \frac{dx_2}{x_2} = -i \lim_{\varepsilon \to 0} \int_{\Delta_\varepsilon} dy \wedge \frac{dx_1}{x_1} \wedge \frac{dx_2}{x_2}.$$

Proof $\int_{\Delta_\varepsilon} dz \wedge \frac{dx_1}{x_1} \wedge \frac{dx_2}{x_2} = - \int_{E'} \log f \, dz \wedge \frac{dg}{g} = 0$ by reason of type.   □

Lemma (8.5)   *Let* $du =$ *either* dx *or* dy. *Then*

$$\lim_{\varepsilon \to 0} \mathrm{Im} \int_{\Phi_\varepsilon} du \wedge \frac{dx_1}{x_1} \wedge \frac{dx_2}{x_2} = -2\pi \, \mathrm{Re} \int_\eta \log(1 - t) \, du.$$

Proof   We have

$$\lim_{\varepsilon \to 0} \mathrm{Im} \int_{\Phi_\varepsilon} du \wedge \frac{dx_1}{x_1} \wedge \frac{dx_2}{x_2}$$

$$= \lim_{\varepsilon \to 0} \mathrm{Im} \int_{T_f} \log f \log g \, du = -2\pi \, \mathrm{Re} \int_{(f)} \log g \, du.$$

The reader can easily reinterpret this integral as the desired integral $\int_\eta$ on $\mathbf{C}^*$.

                                                                             □

Theorem (8.6) *The periods associated to the cycle*

$$\gamma = \gamma_{f.g} = \{(x, f(x), g(x))\} \subset E \times \mathbf{P}^1 \times \mathbf{P}^1$$

are

$$\operatorname{Im} P_\gamma(dx) = \frac{-1}{4\pi^2}\Big(\log|t|\arg(1-t)\Big|_\mathfrak{n} - \operatorname{Im}\int_\mathfrak{n}\log(1-t)\frac{dt}{t}\Big),$$

$$\operatorname{Im} P_\gamma(dy) = \frac{-1}{4\pi^2}\log|t|\log|1-t|\Big|_\mathfrak{n},$$

where P_γ and $\operatorname{Im} P_\gamma$ are as defined in Lecture 3.

Proof Simply use $\frac{dt}{t} = 2\pi i\, dz$ and $y = \frac{-\log|t|}{2\pi}$, together with the equations resulting from Lemmas (8.3)–(8.5):

$$\lim_{\varepsilon\to 0}\operatorname{Im}\int_{\Phi_\varepsilon-\Delta_\varepsilon} dy \wedge \frac{dx_1}{x_1} \wedge \frac{dx_2}{x_2} = -2\pi y\log|1-t|\Big|_\mathfrak{n},$$

$$\lim_{\varepsilon\to 0}\operatorname{Im}\int_{\Phi_\varepsilon-\Delta_\varepsilon} dx \wedge \frac{dx_1}{x_1} \wedge \frac{dx_2}{x_2}$$
$$= \operatorname{Im}\Big(i\Big[2\pi i\, y\log(1-t)\Big|_\mathfrak{n} - 2\pi i\int_\mathfrak{n}\log(1-t)\,dy\Big]\Big),$$

$$-2\pi\operatorname{Re}\int_\mathfrak{n}\log(1-t)\,dx = -2\pi y\arg(1-t)\Big|_\mathfrak{n} - 2\pi\operatorname{Re}\int_\mathfrak{n}\log(1-t)\,dz.$$

□

In order to make things neater, it is convenient to introduce two functions:

$$D(t) = \log|t|\arg(1-t) - \operatorname{Im}\int_0^t\log(1-u)\frac{du}{u},$$

$$J(t) = \log|t|\log|1-t|.$$

We first descend these functions by a process of summation to functions on E.

Lemma (8.7) *The functions*

$$D_q(t) = \sum_{n\in\mathbf{Z}} D(q^n t),$$

$$G_q(t) = \sum_{n=0}^{\infty} J(q^n t) - \sum_{n=1}^{\infty} J(q^n t^{-1})$$
$$+ \frac{(\log|t|)^3}{3\log|q|} - \frac{(\log|t|)^2}{2} + \frac{\log|t|\log|q|}{6}$$

are continuous on $\mathbf{C}^*$ and invariant under $t \to qt$, and hence descend to functions on E.

Proof We refer the reader to Bloch [1] for a proof that the function $D(t)$ is well defined. (At first glance it would appear to depend on a choice of path from 0 to t.) Granting this, one easily checks that $\sum_{n=0}^{\infty} D(tq^n)$ converges. We have

$$\log(1-t)\frac{dt}{t} + \log\left(1 - \frac{1}{t}\right)\frac{d(1/t)}{1/t} = \log(-t)\frac{dt}{t},$$

$$\mathrm{Im}\int^{t} \log(-u)\frac{du}{u} = \mathrm{Im}\left(\frac{1}{2}(\log(-x))^2\right) + C + C'\log|x|,$$

where C' accounts for the ambiguity in the branch of $\log(-t)$. Thus

$$D(t) + D(t^{-1}) = -\frac{1}{2}\mathrm{Im}(\log((-t))^2) + \arg(-t)\log|t| + C + C''\log|t|$$

$$= C + C''\log|t|$$

with C'' accounting for the ambiguity in the branches of $\log(-t)$ and $\arg(1-t)$. For $t < 0$, however, $D(t) = 0$, so for all t we find

(8.7.1) $$D(t) = -D(t^{-1}).$$

From this, one deduces convergence of $\sum_{n<0} D(tq^n)$.

The corresponding continuity and invariance properties of G_q are straightforward, and are left to the reader.   □

Lemma (8.8) *With notation as above,*

$$\mathrm{Im}\,P_\gamma(dx) = \frac{-1}{4\pi^2}D_q(A), \quad \mathrm{Im}\,P_\gamma(dy) = \frac{-1}{4\pi^2}G_q(A),$$

where, writing $(f) = \sum m_j(a_j)$ *and* $(g) = \sum n_k(b_k)$ *with* $a_j, b_k \in E$ *(note change of notation),* $A = \sum m_j n_k(b_j - a_k)$. *(We view D_q and G_q as functions on divisors in the obvious way, for example $D_q(\sum r_i(p_i)) = \sum r_i D_q(p_i)$.)*

Proof The assertion for $\mathrm{Im}\,P_\gamma(dx)$ is immediate from Theorem (8.6). For $\mathrm{Im}\,P_\gamma(dy)$, note that if $\sum n_k(\alpha_k)$ and $\sum m_j(\beta_j)$ are divisors on $\mathbf{C}^*$ with $\sum m_j = \sum n_k = 0$, $\prod \alpha_k^{n_k} = \prod \beta_j^{m_j} = 1$, then the function $(\log|t|)^r$ vanishes on the divisor $\sum m_j n_k(\alpha_k \beta_j^{-1})$ for $r \leq 3$. The reader can now easily check $G_q(A) = \log|t|\log|1 - t|\big|_{\mathrm{n}}$ as claimed.   □

We consider now the complex-valued function

$$Q: \left\{\begin{array}{c} \text{1-cycles on} \\ E \times \mathbf{P}^1 \times \mathbf{P}^1 \end{array}\right\} \to \mathbf{C},$$

(8.8.1)

$$Q(\gamma) = \mathrm{Im}\,P_\gamma(dx) - i\,\mathrm{Im}\,P_\gamma(dy).$$

Notice that Q depends on the choice of a holomorphic differential $dz = dx + i\,dy$ on E. We have

$$Q(\gamma_{f,g}) = \frac{-1}{4\pi^2}(D_q((g) * (g)^-) - iG_q((g) * (f)^-)),$$

where with $(f) = \sum m_j(a_j)$ and $(g) = \sum n_k(b_k)$,

$$g * (f)^- := \sum m_j n_k(b_k - a_j).$$

Recall that given a point of order N on E, which we write as $\frac{k+\ell\tau}{N}$, we have defined a relative algebraic 1-cycle associated to the element $S_{\frac{k+\ell\tau}{N}} \in \Gamma(E, K_2)$

$$S_{\frac{k+\ell\tau}{N}} = \{p, f_{\frac{k+\ell\tau}{N}}\}^N \cdot \prod \{f_i, c_i\}, \quad c_i \in \mathbf{C}^*.$$

Let this cycle be denoted $\gamma_{\frac{k+\ell\tau}{N}}$. As noted earlier it is not completely well defined. However, $Q(\gamma_{\frac{k+\ell\tau}{N}})$ is defined. Our main result can be given a purely analytic formulation as follows: fix $E = C/[1, \tau]$ with $y = \operatorname{Im}\tau > 0$. Let

$$f: \mathbf{Z}/N\mathbf{Z} \times \mathbf{Z}/N\mathbf{Z} \to \mathbf{C}$$

be an odd function, that is $f(a, b) = -f(-a, -b)$, and write

$$\hat{f}(k, \ell) = \frac{1}{N} \sum_{a,b=0}^{N-1} f(a, b)\, e^{2\pi i\left(\frac{-ak+b\ell}{N}\right)}.$$

Theorem (8.9) *With notation as above,*

$$\sum_{k,\ell=0}^{N-1} \hat{f}(k, \ell)Q(\gamma_{\frac{k+\ell\tau}{N}}) = \frac{-y^2N^5}{4\pi^3} \sum_{\substack{m,n\in\mathbf{Z} \\ (m,n)\neq(0,0)}} \frac{f(m, n)}{(m\tau + n)^2(m\bar{\tau} + n)}.$$

Proof As in the first part of the proof of (8.3), one shows that $Q(\gamma_{f,c}) = 0$ for $c \in \mathbf{C}^*$, whence

$$(8.9.1) \qquad Q(\gamma_{\frac{k+\ell\tau}{N}}) = \frac{-N}{4\pi^2}\left(D_q((\rho)^-_*(f_{\frac{k+\ell\tau}{N}})) - iG_q((\rho)^-_*(f_{\frac{k+\ell\tau}{N}}))\right).$$

Writing $Q_q = \frac{-1}{4\pi^2}(D_q - iG_q)$, viewed as a function on E, one has

$$Q(\gamma_{\frac{k+\ell\tau}{N}}) = N^2 \sum_{\substack{Na=0 \\ a\neq 0}} \left(Q_q(a) - Q_q\left(a + \frac{k+\ell\tau}{N}\right)\right)$$

$$+ N^2(N^2 - 1)\left(Q_q\left(\frac{k+\ell\tau}{N}\right) - Q_q(0)\right).$$

Since $Q_q(0) = 0$, this yields

$$Q(\gamma_{\frac{k+\ell\tau}{N}}) = N^4 Q_q\left(\frac{k+\ell\tau}{N}\right).$$

Lemma (8.10)

$$\sum_{k,\ell=0}^{N-1} \hat{f}(k,\ell)\left(\frac{\ell^3}{3N^3} - \frac{\ell^2}{2N^2} + \frac{\ell}{6N}\right) = \frac{iN}{4\pi^3} \sum_{\substack{n\in\mathbf{Z}\\n\neq 0}} \frac{f(0,n)}{n^3}.$$

Proof We use the well-known Fourier series

$$\sum_{\substack{n\in\mathbf{Z}\\n\neq 0}} \frac{e^{2\pi i n x}}{n^3} = 4\pi^3 i\left(\frac{x^3}{3} - \frac{x^2}{2} + \frac{x}{6}\right)$$

to write

$$\sum_{k,\ell} \hat{f}(k,\ell)\left(\frac{\ell^3}{3N^3} - \frac{\ell^2}{2N^2} + \frac{\ell}{6N}\right)$$

$$= \frac{-i}{4\pi^3} \sum_{k,\ell} \hat{f}(k,\ell) \sum_{\substack{n\in\mathbf{Z}\\n\neq 0}} \frac{e^{2\pi i \ell n/N}}{n^3}$$

$$= \frac{-i}{4\pi^3 N} \sum_{a,b,k,\ell} f(a,b)\, e^{2\pi i\left(\frac{-ak+b\ell}{N}\right)} \sum_{n\neq 0} \frac{e^{2\pi i \ell n/N}}{n^3}$$

$$= \frac{-iN}{4\pi^3} \sum_{n\neq 0} \frac{f(0,-n)}{n^3} = \frac{iN}{4\pi^3} \sum \frac{f(0,n)}{n^3}. \qquad \square$$

It is convenient now to split up the function Q_q. For this, we view Q_q, D_q, G_q as functions on $\mathbf{C}^*$, and write

$$Q_q(t) = \frac{-1}{4\pi^2}(D_q(t) - iG_q(t))$$

$$= \frac{i}{4\pi^2}\left(\sum_{n=0}^{\infty} \log|tq^n|\log(1 - tq^n) - \sum_{n=1}^{\infty} \log|t^{-1}q^n|\log(1 - t^{-1}q^n)\right)$$

$$+ \frac{1}{4\pi^2}\left(\sum_{n=0}^{\infty} \mathrm{Im}\int_{x^{-1}q^n}^{xq^n} \log(1 - t)\frac{dt}{t}\right)$$

$$+ \frac{i}{4\pi^2}\left(\frac{(\log|t|)^3}{3\log|q|} - \frac{(\log|t|)^2}{2} + \frac{\log|t|\log|q|}{6}\right).$$

(In what follows $|xq^n| < 1$ for $n \geq 0$ and $|x^{-1}q^n| < 1$ for $n \geq 1$, so there will be no ambiguity about branches of functions.) Write the three terms in parentheses as $R_q'(t)$, $R_q''(t)$, and $R_q'''(t)$ respectively, so

$$Q_q(t) = \frac{i}{4\pi^2}R_q'(t) + \frac{1}{4\pi^2}R_q''(t) + \frac{i}{4\pi^2}R_q'''(t).$$

We have already seen

$$\frac{i}{4\pi^2} \sum_{k,\ell=0}^{N-1} \hat{f}(k,\ell)\, R_q''' \left(e^{2\pi i \left(\frac{k+\ell\tau}{N} \right)} \right)$$

$$= \frac{i}{4\pi^2} \sum_{k,\ell=0}^{N-1} \hat{f}(k,\ell)\left(\frac{4\pi^2 y^2 \ell^3}{3N^3} - \frac{4\pi^2 y^2 \ell^2}{2N^2} + \frac{4\pi^2 y^2 \ell}{6N} \right)$$

$$= iy^2 \sum_{k,\ell} \hat{f}(k,\ell)\left(\frac{\ell^3}{3N^3} - \frac{\ell^2}{2N^2} + \frac{\ell}{6N} \right) = \frac{-Ny^2}{4\pi^3} \sum_{\substack{n\in\mathbb{Z} \\ n\neq 0}} \frac{f(0,n)}{n^3}.$$

The formula in (8.9) to be proven now reads

$$(8.11) \qquad \sum_{k,\ell=0}^{N-1} \hat{f}(k,\ell)\left(R_q'' \left(e^{2\pi i \left(\frac{k+\ell\tau}{N} \right)} \right) + iR' \left(e^{2\pi i \left(\frac{k+\ell\tau}{N} \right)} \right) \right)$$

$$= \frac{-y^2 N}{\pi} \sum_{\substack{m,n\in\mathbb{Z} \\ m\neq 0}} \frac{f(m,n)}{(m\tau + n)^2 (m\bar{\tau} + n)}.$$

Put

$$(8.12) \qquad L = \frac{1}{N} \sum_{k,\ell=0}^{N-1} \hat{f}(k,\ell)\, R_q' \left(e^{2\pi i \left(\frac{k+\ell\tau}{N} \right)} \right),$$

$$M = \frac{-i}{N} \sum \hat{f}(k,\ell)\, R_q'' \left(e^{2\pi i \left(\frac{k+\ell\tau}{N} \right)} \right).$$

Proposition (8.13) *We have*

$$L = \frac{-y}{2\pi} \sum_{\substack{m\in\mathbb{Z} \\ m\neq 0}} \sum_{n\in\mathbb{Z}} \frac{f(m,n)}{m(m\tau + n)^2}.$$

Proof

$$L = \frac{-2\pi y}{N} \sum_{k,\ell=0}^{N-1} \hat{f}(k,\ell)\left[\sum_{n=0}^{\infty} \left(n + \frac{\ell}{N} \right) \log\left(1 - e^{2\pi i \left(\frac{k+\ell\tau}{N} + n\tau \right)} \right) \right.$$

$$\left. - \sum_{n-1}^{\infty} \left(n - \frac{\ell}{N} \right) \log\left(1 - e^{2\pi i \left(\frac{-k-\ell\tau}{N} + n\tau \right)} \right) \right].$$

To consolidate the two series, note that (replacing k, ℓ by $N - k$, $N - \ell$, and n

by $n + 1$),

$$- \sum_{k,\ell=0}^{N-1} \hat{f}(k,\ell) \sum_{n=1}^{\infty} \left(n - \frac{\ell}{N}\right) \log\left(1 - e^{2\pi i\left(\frac{-k-\ell\tau}{N}+n\tau\right)}\right)$$

$$= \sum_{k=0}^{N-1} \sum_{\ell=1}^{N} \hat{f}(k,\ell) \sum_{n=0}^{\infty} \left(n + \frac{\ell}{N}\right) \log\left(1 - e^{2\pi i\left(\frac{k+\ell\tau}{N}+n\tau\right)}\right)$$

$$= \sum_{k,\ell=0}^{N-1} \hat{f}(k,\ell) \sum_{n=0}^{\infty} \left(n + \frac{\ell}{N}\right) \log\left(1 - e^{2\pi i\left(\frac{k+\ell\tau}{N}+n\tau\right)}\right)$$

$$+ \sum_{k=0}^{N-1} \hat{f}(k,N) \sum_{n=1}^{\infty} n \log\left(1 - e^{2\pi i\left(\frac{k}{N}+n\tau\right)}\right)$$

$$- \sum_{k=0}^{N-1} \hat{f}(k,0) \sum_{n=0}^{\infty} n \log\left(1 - e^{2\pi i\left(\frac{k}{N}+n\tau\right)}\right)$$

$$= \sum_{k,\ell=0}^{N-1} \hat{f}(k,\ell) \sum_{n=0}^{\infty} \left(n + \frac{\ell}{N}\right) \log\left(1 - e^{2\pi i\left(\frac{k+\ell\tau}{N}+n\tau\right)}\right).$$

Thus

$$L = \frac{-4\pi y}{N} \sum_{k,\ell=0}^{N-1} \hat{f}(k,\ell) \sum_{n=0}^{\infty} \left(n + \frac{\ell}{N}\right) \log\left(1 - e^{2\pi i\left(\frac{k+\ell\tau}{N}+n\tau\right)}\right)$$

$$= \frac{-4\pi y}{N^2} \sum_{k,\ell,a,b} f(a,b) \, e^{2\pi i\left(\frac{-ak+b\ell}{N}\right)} \sum_{n=0}^{\infty} \sum_{m=1}^{\infty} \left(n + \frac{\ell}{N}\right) \frac{e^{2\pi i m\left(\frac{k+\ell\tau}{N}+n\tau\right)}}{m}.$$

This can be rewritten

(8.14) $$\qquad L = \frac{4\pi y}{N^2} \sum_{a,b} f(a,b) \sum_{\substack{n=0 \\ m=1 \\ m \equiv a (\mathrm{mod}\, N)}}^{\infty} \frac{n e^{2\pi i\left(\frac{m\tau+b}{N}\right)}}{m}.$$

Lemma (8.15)   *For* Im $x > 0$,

$$\sum_{n=0}^{\infty} n e^{2\pi i n x} = \frac{-1}{4 \sin^2 \pi x}.$$

Proof of lemma

$$\sum n e^{2\pi i n x} = \frac{1}{2\pi i} \frac{d}{dx} \left(\frac{1}{1 - e^{2\pi i x}}\right) = \frac{e^{2\pi i x}}{(1 - e^{2\pi i x})^2}$$

$$= \frac{1}{(e^{\pi i x} - e^{-\pi i x})^2} = \frac{-1}{4 \sin^2 \pi x}. \qquad \square$$

Returning to the expression for L (8.14), we see

$$L = \frac{-\pi y}{N^2} \sum_{a,b} f(a,b) \sum_{\substack{m=1 \\ m \equiv a (\text{mod } N)}}^{\infty} \frac{1}{m \sin^2 \pi(\frac{m\tau + b}{N})}.$$

Now use

$$\frac{\pi^2}{\sin^2 \pi x} = \sum_{n \in \mathbf{Z}} \frac{1}{(x+n)^2}$$

to write

$$L = \frac{-y}{\pi N^2} \sum_{a,b} f(a,b) \sum_{\substack{m=1 \\ m \equiv a (\text{mod } N)}}^{\infty} \sum_{n \in \mathbf{Z}} \frac{1}{m(\frac{m\tau + b}{N} + n)^2}$$

$$= \frac{-y}{2\pi} \sum_{\substack{m,n=-\infty \\ m \neq 0}}^{\infty} \frac{f(m,n)}{m(m\tau + n)^2}.$$

This proves (8.13). □

Proposition (8.16) *Let M be as in (8.12). Then*

$$M = \frac{-1}{2\pi} \sum_{\substack{m,n=-\infty \\ m \neq 0}}^{\infty} f(m,n) \, \text{Im}\left(\frac{1}{m^2(m\tau + n)}\right).$$

Proof For $|x| < 1$, the "natural" branch of $\int_0^x \log(1-t)\frac{dt}{t}$ is $-\sum_{m=1}^{\infty} \frac{x^m}{m^2}$. Thus

$$M = \frac{i}{N} \sum_{k,\ell=0}^{N-1} \hat{f}(k,\ell) \, \text{Im}\left(\sum_{\substack{n=0 \\ m=1}}^{\infty} \frac{e^{2\pi i\left(\frac{k+\ell\tau}{N} + n\tau\right)m}}{m^2} - \sum_{\substack{n=1 \\ m=1}}^{\infty} \frac{e^{2\pi i\left(\frac{-k-\ell\tau}{N} + n\tau\right)m}}{m^2} \right).$$

As with L, it will be convenient to combine the two sums:

$$-\sum_{k,\ell=0}^{N-1} \hat{f}(k,\ell) \sum_{m=n=1}^{\infty} \frac{e^{2\pi i m\left(\frac{k+\ell\tau}{N} + n\tau\right)}}{m^2}$$

$$= \sum_{k=0}^{N-1} \sum_{\ell=1}^{N} \hat{f}(k,\ell) \sum_{\substack{m=1 \\ n=0}}^{\infty} \frac{e^{2\pi i m\left(\frac{k+\ell\tau}{N} + n\tau\right)}}{m^2}$$

$$= \sum_{k,\ell=0}^{N-1} \hat{f}(k,\ell) \sum_{\substack{m=1 \\ n=0}}^{\infty} \frac{e^{2\pi i m\left(\frac{k+\ell\tau}{N} + n\tau\right)}}{m^2} - \sum_{k=0}^{N-1} \hat{f}(k,0) \sum_{m=1}^{\infty} \frac{e^{2\pi i \frac{mk}{N}}}{m^2},$$

so

$$M = \frac{2i}{N} \sum_{k,\ell=0}^{N-1} \hat{f}(k,\ell) \, \mathrm{Im}\left(\sum_{\substack{n=0 \\ m=1}}^{\infty} \frac{e^{2\pi i \left(\frac{k+\ell\tau}{N} + n\tau \right) m}}{m^2} \right)$$

$$- \frac{i}{N} \sum_{k=0}^{N-1} \hat{f}(k,0) \, \mathrm{Im}\left(\sum_{m=1}^{\infty} \frac{e^{2\pi i \frac{mk}{N}}}{m^2} \right) = M_1 + M_2.$$

Lemma (8.17)

$$M_1 = \frac{1}{N} \sum_{b=0}^{N-1} \sum_{m=1}^{\infty} \frac{f(m,b)}{m^2} - \frac{1}{2\pi} \sum_{\substack{m,n=-\infty \\ m\neq 0}}^{\infty} f(m,n) \, \mathrm{Im}\left(\frac{1}{m^2(m\tau+n)} \right).$$

Proof

$$M_1 = \frac{2i}{N^2} \sum_{a,b,k,\ell} f(a,b) \, e^{2\pi i \left(\frac{-ak+b\ell}{N} \right)} \mathrm{Im}\left(\sum_{\substack{n=0 \\ m=1}}^{\infty} \frac{e^{2\pi i \left(\frac{k+\ell\tau}{N} + n\tau \right) m}}{m^2} \right).$$

Note that $\sum_{a,b} f(a,b) \, \mathrm{Re}(e^{2\pi i \left(\frac{-ak+b\ell}{N} \right)}) = 0$ because $f(a,b) = -f(-a,-b)$, and also

$$\mathrm{Im}\left(i e^{2\pi i \left(\frac{-ak+b\ell}{N} \right)}(*) \right) = -\mathrm{Im}\left(e^{2\pi i \left(\frac{-ak+b\ell}{N} \right)} \right) \mathrm{Im}(*) + \mathrm{Re}\left(e^{2\pi i \left(\frac{-ak+b\ell}{N} \right)} \right) \mathrm{Re}(*).$$

It follows that

$$M_1 = \frac{2}{N^2} \sum_{a,b,k,\ell} f(a,b) \, \mathrm{Im}\left(i \sum_{\substack{n=0 \\ m=1}}^{\infty} \frac{e^{2\pi i \left(\frac{(m-a)k}{N} + (nN+\ell)\frac{m\tau+b}{N} \right)}}{m^2} \right)$$

$$= \frac{2}{N} \sum_{a,b} f(a,b) \, \mathrm{Im}\left(i \sum_{\substack{m=1 \\ m\equiv a(\mathrm{mod}\, N)}}^{\infty} \sum_{n=0}^{\infty} \frac{e^{2\pi i n \left(\frac{m\tau+b}{N} \right)}}{m^2} \right).$$

Using the identity

$$\frac{1}{1 - e^{2\pi i x}} = \frac{i \cot \pi x + 1}{2},$$

we can write

$$M_1 = \frac{1}{N} \sum_{a,b} f(a,b) \, \mathrm{Im}\left(\sum_{\substack{m=1 \\ m\equiv a(\mathrm{mod}\, N)}}^{\infty} \frac{-\cot \pi\left(\frac{m\tau+b}{N} \right) + i}{m^2} \right)$$

$$= \frac{1}{N} \sum_{b} \sum_{m=1}^{\infty} \frac{f(m,b)}{m^2} - \frac{1}{N} \sum_{a,b} f(a,b) \, \mathrm{Im}\left(\sum_{\substack{m=1 \\ m\equiv a}}^{\infty} \frac{1}{m^2} \cot\left[\pi\left(\frac{m\tau+b}{N} \right) \right] \right).$$

Note that

$$\cot \pi x = \frac{1}{\pi} \lim_{N\to\infty} \sum_{n=-N}^{N} \frac{1}{x+n},$$

and

$$\mathrm{Im}(\cot \pi x) = \frac{1}{\pi} \sum_{n\in\mathbf{Z}} \mathrm{Im}\left(\frac{1}{x+n}\right)$$

(the series converging without any special sort of summation), so we get

$$M_1 = \frac{1}{N} \sum_{\substack{m=1\\b}}^{\infty} \frac{f(m,b)}{m^2} - \frac{1}{\pi N} \sum_{a,b} f(a,b) \sum_{\substack{m=1\\m\equiv a}}^{\infty} \sum_{n\in\mathbf{Z}} \mathrm{Im}\left(\frac{1}{m^2(\frac{m\tau+b}{N}+n)}\right)$$

$$= \frac{1}{N} \sum_{\substack{m=1\\b}}^{\infty} \frac{f(m,b)}{m^2} - \frac{1}{2\pi} \sum_{\substack{m\in\mathbf{Z}\\m\neq 0}} \sum_{n\in\mathbf{Z}} f(m,n) \mathrm{Im}\left(\frac{1}{m^2(m\tau+n)}\right).$$

This completes the computation of M_1. □

Lemma (8.18) $\quad M_2 = \dfrac{-1}{N} \sum_{\substack{m=1\\b=0}}^{\substack{b=N-1\\m=\infty}} \dfrac{f(m,b)}{m^2}.$

Proof

$$M_2 = \frac{-i}{N} \sum_{k=0}^{N-1} \hat{f}(k,0) \mathrm{Im}\left(\sum_{m=1}^{\infty} \frac{e^{2\pi i\frac{mk}{N}}}{m^2}\right)$$

$$= \frac{-i}{N^2} \sum_{k,a,b} f(a,b) e^{\frac{-2\pi iak}{N}} \mathrm{Im}\left(\sum_{m=1}^{\infty} \frac{e^{2\pi i\frac{mk}{N}}}{m^2}\right)$$

$$= \frac{-1}{N^2} \sum_{k,a,b} f(a,b) \mathrm{Im}\left(i \sum_{m=1}^{\infty} \frac{e^{2\pi i\frac{k}{N}(m-a)}}{m^2}\right)$$

$$= \frac{-1}{N} \sum \frac{f(m,b)}{m^2}. \qquad \square$$

We return now to the proof of (8.16).

$$M = M_1 + M_2 = \frac{-1}{2\pi} \sum_{m\neq 0} \sum_{n\in\mathbf{Z}} f(m,n) \mathrm{Im}\left(\frac{1}{m^2(m\tau+n)}\right)$$

as desired. □

We can now complete the proof of Theorem (8.9):

$$\sum_{k,\ell=0}^{N-1} \hat{f}(k,\ell)\left(R_q''\left(e^{2\pi i\left(\frac{k+\ell\tau}{N}\right)}\right) + i R_q'\left(e^{2\pi i\left(\frac{k+\ell\tau}{N}\right)}\right)\right) = i\,N(L+M)$$

$$= \frac{-i\,yN}{2\pi}\Big(\sum_{\substack{m\in\mathbf{Z}\\ m\neq 0}}\sum_{n\in\mathbf{Z}} f(m,n)\big(\frac{1}{m(m\tau+n)^2} + \frac{1}{y}\,\mathrm{Im}\frac{1}{m^2(m\tau+n)}\big)\Big).$$

But

$$\frac{1}{m(m\tau+n)^2} + \frac{1}{y}\,\mathrm{Im}\Big(\frac{1}{m^2(m\tau+n)}\Big)$$

$$= \frac{1}{m(m\tau+n)^2} + \frac{1}{2iy}\left(\frac{1}{m^2\,(m\tau+n)} - \frac{1}{m^2\,(m\bar{\tau}+n)}\right)$$

$$= \frac{1}{m(m\tau+n)^2} - \frac{1}{m(m\tau+n)(m\bar{\tau}+n)}$$

$$= -2i\,y\frac{1}{(m\tau+n)^2\,(m\bar{\tau}+n)},$$

so

$$\sum_{k,\ell=0}^{N-1} \hat{f}(k,\ell)\left(R_q''(e^{2\pi i\frac{k+\ell\tau}{N}}) + iR'(e^{2\pi i\frac{k+\ell\tau}{N}})\right) = \frac{-y^2 N}{\pi}\sum_{\substack{m,n\in\mathbf{Z}\\ m\neq 0}} \frac{f(m,n)}{(m\tau+n)^2\,(m\bar{\tau}+n)}.$$

This completes the proof of (8.11) and also (8.9).   □

References for Lecture 8

[1] S. Bloch, Applications of the dilogarithm function in algebraic K-theory and algebraic geometry, pp. 103–114 in *Proceedings of the International Symposium on Algebraic Geometry (Kyoto Univ., Kyoto, 1977)*, Kinokuniyo Book Store, Tokyo (1978).

[2] S. Bloch, *Higher Regulators, Algebraic K-Theory, and Zeta Functions of Elliptic Curves*, Lectures given at the University of California, Irvine (1978). [CRM Monograph Series, no. 11, American Mathematical Society, Providence, R.I., 2000.]

[3] A. Borel, Cohomologie de SL$_n$ et valeurs de fonctions zeta aux points entiers, *Ann. Scuola Norm. Sup. Pisa Cl. Sci. (4)*, **4** (1977), 613–636; errata, **7** (1980), 373.

[4] S. Lang, *Elliptic Functions*, Addison-Wesley, Reading, Mass. (1973). [Second edition: Springer, New York, 1987.]

9

Relative cycles and zeta functions – continued

In this final lecture we will show how Theorem (8.9) leads to a regulator formula for the value at $s = 2$ of the zeta function of an elliptic curve E with complex multiplication by the ring of integers in an imaginary quadratic field k with class number 1. We begin with some general lemmas, valid without hypotheses on E. Notation will be as in Lecture 8.

Lemma (9.1) $Q(\gamma_{\frac{a+b\tau}{N}}) = -Q(\gamma_{\frac{-a-b\tau}{N}}).$

Proof Using formulas (8.7.1) and (8.9.1), together with the fact that the divisor (ρ) is invariant under $z \to -z$, we reduce to proving $G_q(t^{-1}) = -G_q(t)$. We have

$$G_q(t^{-1}) + G_q(t) = \sum_{n=0}^{\infty} J(q^n t^{-1}) - \sum_{n=1}^{\infty} J(q^n t)$$

$$+ \sum_{n=0}^{\infty} J(q^n t) - \sum_{n=1}^{\infty} J(q^n t^{-1}) - (\log|t|)^2$$

$$= J(t^{-1}) + J(t) - (\log|t|)^2 = 0. \qquad \square$$

Lemma (9.2) $Q(\gamma_{\frac{a+b}{N}}) = \dfrac{i y^2 N^4}{4\pi^3} \sum_{(m,n)\neq(0,0)} \dfrac{\sin(2\pi(\frac{an-bm}{N}))}{(m+n\tau)^2 (m+n\bar{\tau})}.$

Proof   Let $g_{a+b\tau}(k + \ell\tau) = \frac{1}{N}e^{2\pi i\left(\frac{a\ell-bk}{N}\right)}$. Then

$$\hat{g}_{a+b\tau}(m + n\tau) = \frac{1}{N^2}\sum_{k,\ell}e^{2\pi i\left(\frac{a\ell-bk}{N}\right)}e^{2\pi i\left(\frac{-kn+\ell m}{N}\right)}$$

$$= \frac{1}{N^2}\sum_{k,\ell}e^{2\pi i\left(\frac{\ell(a+m)-k(b+n)}{N}\right)}$$

$$= \begin{cases} 0 & a \neq -m \quad \text{or} \quad b \neq -n \\ 1 & a = -m \quad \text{or} \quad b = -n \end{cases}$$

$$= \delta_{-a-b\tau}(m + n\tau)$$

with obvious notation. Writing $f_{a+b\tau} = g_{-a-b\tau} - g_{a+b\tau}$ we get

$$Q(\gamma_{\frac{a+b\tau}{N}}) = \frac{1}{2}\sum \hat{f}_{a+b\tau}(k + \ell\tau)Q(\gamma_{\frac{k+\ell\tau}{N}}) = \frac{-y^2 N^5}{8\pi^3}\frac{f_{a+b\tau}(m + n\tau)}{(m + n\tau)^2(m + n\bar{\tau})}$$

$$= \frac{i\,y^2 N^4}{4\pi^3}\sum\frac{\sin\left(2\pi\frac{(an-bm)}{N}\right)}{(m + n\tau)^2(m + n\bar{\tau})}. \qquad\qquad \square$$

The following description of the "modular behavior" of R_q seemed worth including although it will not be used:

Lemma (9.3)   *Let* $\tau' = \frac{\alpha\tau+\beta}{\gamma\tau+\delta}$ *with* $\left(\begin{smallmatrix}\alpha & \gamma \\ \beta & \delta\end{smallmatrix}\right) \in SL_2(\mathbf{Z})$. *Write* $q' = e^{2\pi i\tau'}$ *and let* $\left(\begin{smallmatrix}b \\ a\end{smallmatrix}\right) = \left(\begin{smallmatrix}\alpha & \gamma \\ \beta & \delta\end{smallmatrix}\right)\left(\begin{smallmatrix}b' \\ a'\end{smallmatrix}\right)$. *Then*

$$Q_{q'}(\gamma_{\frac{a'+b'\tau'}{N}}) = (\gamma\bar{\tau} + \delta)^{-1}Q_q(\gamma_{\frac{a+b\tau}{N}}).$$

(We write Q_q and $Q_{q'}$ to indicate dependence on q and q'.)

Proof   Since $SL_2(\mathbf{Z})$ preserves the symplectic form $\left(\begin{smallmatrix}0 & 1 \\ -1 & 0\end{smallmatrix}\right)$, we find that $-a'\ell' + b'k' = -a\ell + bk$, where $\left(\begin{smallmatrix}\ell \\ k\end{smallmatrix}\right) = \left(\begin{smallmatrix}\alpha & \gamma \\ \beta & \delta\end{smallmatrix}\right)\left(\begin{smallmatrix}\ell' \\ k'\end{smallmatrix}\right)$. Thus

$$Q_{q'}(\gamma_{\frac{a'+b'\tau'}{N}}) = \frac{i\,y'^2 N^4}{4\pi^3}\sum_{k',\ell'}\frac{\sin\left(2\pi\frac{a'\ell'-b'k'}{N}\right)}{(k' + \ell'\tau')^2(k' + \ell'\bar{\tau}')}$$

$$= \frac{i\,y'^2 N^4}{4\pi^3}\sum_{k',\ell'}\frac{\sin\left(2\pi\frac{a\ell-bk}{N}\right)}{(k' + \ell'\tau')^2(k' + \ell'\bar{\tau}')}.$$

Note that

$$k' + \ell'\tau' = (\gamma\tau + \delta)^{-1}(k'\delta + \ell'\beta + (k'\gamma + \ell'\alpha)\tau)$$

$$= (\gamma\tau + \delta)^{-1}(k + \ell\tau),$$

whence (using $y' = \frac{y}{(\gamma\tau+\delta)(\gamma\bar{\tau}+\delta)}$)

$$Q_{q'}(\gamma_{\frac{a'+b'\tau'}{N}}) = \frac{i\,y^2 N^4 (\gamma\tau+\delta)^2(\gamma\bar{\tau}+\delta)}{4\pi^3(\gamma\tau+\delta)^2(\gamma\bar{\tau}+\delta)^2} \sum_{k,\ell} \frac{\sin\left(2\pi\,\frac{a\ell-bk}{N}\right)}{(k+\ell\tau)^2(k+\ell\bar{\tau})}$$

$$= (\gamma\bar{\tau}+\delta)^{-1} Q_q(\gamma_{\frac{a+b\tau}{N}}).$$ □

It is convenient to denote by $\langle\ \rangle$ the bilinear form

$$\langle\ \rangle : \mathfrak{o}/N\mathfrak{o} \times \mathfrak{o}/N\mathfrak{o} \to \mathbf{C}^*,$$

$$\langle a+b\tau, k+\ell\tau \rangle = e^{2\pi i\left(\frac{-a\ell+bk}{N}\right)}.$$

We assume henceforth that $\mathfrak{o} = \mathbf{Z} + \mathbf{Z}\tau$ is an order in an imaginary quadratic field κ.

Lemma (9.4) $\langle xy, z \rangle = \langle x, \bar{y}z \rangle$, *where $\bar{y}$ is the complex conjugate.*

Proof   Write $x = x_1+x_2\tau$, $y = y_1+y_2\tau$, $z = z_1+z_2\tau$, and suppose $\tau^2+A\tau+B = 0$, with $x_i, y_i, z_i, A, B \in \mathbf{Z}$. Then

$$xy = x_1 y_1 - x_2 y_2 B + (x_1 y_2 + x_2 y_1 - x_2 y_2 A)\tau,$$

$$\langle xy, z \rangle = \exp\left(2\pi i\,\frac{x_2 y_2 B z_2 - x_1 y_1 z_2 + x_1 y_2 z_1 + x_2 y_1 z_1 - x_2 y_2 A z_1}{N}\right),$$

$$\bar{y}z = y_1 z_1 + y_2 z_2 B + y_2 z_1 \bar{\tau} + y_1 z_2 \tau$$

$$= (y_1 z_1 + y_2 z_2 B - y_2 z_1 A) + (y_1 z_2 - y_2 z_1)\tau,$$

$$\langle x, \bar{y}z \rangle = \exp\left(2\pi i\,\frac{-x_1 y_1 z_2 + x_1 y_2 z_1 + x_2 y_1 z_1 + x_2 y_2 z_2 B - x_2 y_2 z_1 A}{N}\right).$$ □

Corollary (9.5)   *If $\zeta \in \mathfrak{o}_\kappa$ is a unit (i.e. a root of 1), then $Q(\gamma_{\frac{\zeta(a+b\tau)}{N}}) = \zeta^{-1} Q(\gamma_{\frac{a+b\tau}{N}})$.*

Proof

$$Q(\gamma_{\frac{\zeta(a+b\tau)}{N}}) = \frac{i\,y^2 N^4}{4\pi^3} \frac{\mathrm{Im}\langle \zeta(a+b\tau), k+\ell\tau \rangle}{(k+\ell\tau)^2(k+\ell\bar{\tau})}$$

$$= \frac{i\,y^2 N^4}{4\pi^3} \sum \frac{\mathrm{Im}\langle a+b\tau, \bar{\zeta}(k+\ell\tau) \rangle}{(k+\ell\tau)^2(k+\ell\bar{\tau})}$$

$$= \frac{i\,y^2 N^4}{4\pi^3} \sum \frac{\mathrm{Im}\langle a+b\tau, k'+\ell'\tau \rangle}{(k'+\ell'\tau)^2(k'+\ell'\bar{\tau})}$$

$$= \zeta^{-1} Q(\gamma_{\frac{a+b\tau}{N}}).$$ □

Assume now for simplicity that $\mathfrak{o}$ is the ring of integers in κ, and κ has class number 1.

Corollary (9.6) *Suppose $N = fg$, with $f, g \in \mathfrak{o}_k$. Then*

$$\bar{g} \sum_{\mu \in \mathfrak{o}/g\mathfrak{o}} Q(\gamma_{\frac{a+b\tau}{N} + \frac{f\mu}{N}}) = Q(\gamma_{\frac{a+b\tau}{f}}).$$

Proof We have

$$\sum_{\mu \in \mathfrak{o}/g\mathfrak{o}} \langle \mu, \bar{f}(k + \ell\tau)\rangle = \begin{cases} g\bar{g} & \bar{g} \mid k + \ell\tau, \\ 0 & \text{otherwise.} \end{cases}$$

Thus

$$
\begin{aligned}
\sum_{\mu \in \mathfrak{o}/g\mathfrak{o}} Q(\gamma_{\frac{a+b\tau}{N} + \frac{f\mu}{N}}) &= \sum_{k,\ell} \frac{i\,y^2 N^4}{4\pi^3} \sum_{\mu} \langle \mu, \bar{f}(k + \ell\tau)\rangle \frac{\mathrm{Im}\langle a + b\tau, k + \ell\tau\rangle}{(k + \ell\tau)^2(k + \ell\bar{\tau})} \\
&= \frac{i\,g\bar{g}y^2 N^4}{4\pi^3} \sum_{r,s} \frac{\mathrm{Im}\langle a + b\tau, \bar{g}(r + s\tau)\rangle}{(r + s\tau)^2(r + s\bar{\tau})\bar{g}^2 g} \\
&= \frac{i\,y^2 N^4}{4\bar{g}\pi^3} \sum_{r,s} \frac{\mathrm{Im}\langle g(a + b\tau), r + s\tau\rangle}{(r + s\tau)^2(r + s\bar{\tau})} \\
&= (\bar{g})^{-1} Q(\gamma_{\frac{g(a+b\tau)}{N}}) \\
&= (\bar{g})^{-1} Q(\gamma_{\frac{a+b\tau}{f}}).
\end{aligned}
$$

$\square$

Lemma (9.7) *Let χ have conductor $f \mid N$, $n = fg$. Then*

(i) $\hat{\chi}(x) = 0$ unless $N \mid \bar{f}x$.
(ii) For $x \in (\mathfrak{o}/N\mathfrak{o})^*$, $\hat{\chi}(xy) = \bar{\chi}(\bar{x})\chi(y)$.
(iii) Let $f_1 \mid f$, $f \nmid f_1$, and suppose $N \mid \bar{f}_1 x$. Then $\hat{\chi}(x) = 0$.

Proof (i) We have

$$
\begin{aligned}
\hat{\chi}(x) &= \frac{1}{N} \sum_{y \in \mathfrak{o}/N\mathfrak{o}} \chi(y)\langle x, y\rangle = \frac{1}{N} \sum_{y' \bmod f} \chi(y') \sum_{y'' \bmod g} \langle x, y'\rangle\langle x, fy''\rangle \\
&= \frac{1}{N} \sum_{y' \bmod f} \chi(y')\langle x, y'\rangle \sum_{y'' \bmod g} \langle x\bar{f}, y''\rangle \\
&= \begin{cases} 0 & \bar{g} \nmid x, \\ \frac{g\bar{g}}{N} \sum_{y' \bmod f} \chi(y')\langle x, y'\rangle & \bar{g} \mid x, \end{cases}
\end{aligned}
$$

where the y' (resp. y'') run through representatives in $\mathfrak{o}$ for the congruence classes modulo f (resp. g).

(ii) If $x \in (\mathfrak{o}/N\mathfrak{o})^*$,

$$\hat{\chi}(xy) = \frac{1}{N} \sum_{z \in \mathfrak{o}/N\mathfrak{o}} \chi(z)\langle xy, z\rangle = \frac{1}{N} \bar{\chi}(\bar{y}) \sum_z \chi(z\bar{y})\langle x, \bar{y}z\rangle = \bar{\chi}(\bar{y})\hat{\chi}(x).$$

(iii) Since conductor $\chi = f \nmid f_1$, we can find $y \equiv 1 \mod \bar{f}_1$, y a unit in $\mathfrak{o}/N\mathfrak{o}$, with $\chi(\bar{y}) \neq 1$. Since $N \mid \bar{f}_1 \cdot x$

$$\hat{\chi}(x) = \hat{\chi}(xy) = \bar{\chi}(\bar{y})\hat{\chi}(x).$$

Thus $\hat{\chi}(x) = 0$. $\qquad\qquad\qquad\qquad\qquad\qquad\qquad\qquad\qquad\square$

We now have the tools we need for the regulator formula. Assume as above that $\mathfrak{o} = \mathbf{Z} + \mathbf{Z}\tau$ is the ring of integers in an imaginary quadratic field κ of class number 1. We fix an embedding $\kappa \to \mathbf{C}$, an integer N, and a character χ of $(\mathfrak{o}/N\mathfrak{o})^*$ which restricts to the given embedding $\mu_\kappa \to \mathbf{C}^*$ on the roots of 1. Let $\chi^{\text{Gröss}}$ denote the Grössencharakter

$$\chi^{\text{Gröss}}(\mathfrak{p}) = \bar{h}\chi(h), \qquad \mathfrak{p} = (h), \quad \mathfrak{p} \nmid N.$$

Let f generate the conductor ideal of χ and write $N = fg$. From (8.9), (9.5), (9.6), and (9.7) we get

$$(9.8) \quad L(2, \chi^{\text{Gröss}}) = \frac{4\pi^3}{-y^2 N^5} \sum_{w \in \mathfrak{o}/N\mathfrak{o}} \hat{\chi}(w) Q(\gamma_{\frac{w}{N}})$$

$$= \frac{4\pi^3}{-y^2 N^5} \sum_{x \in (\mathfrak{o}/\bar{f}\mathfrak{o})^*} \hat{\chi}(x\bar{g}) Q(\gamma_{x\bar{f}^{-1}}) = \frac{4\pi^3 \hat{\chi}(\bar{g})}{-y^2 N^5} \sum_{x \in (\mathfrak{o}/\bar{f}\mathfrak{o})^*} \bar{\chi}(\bar{x}) Q(\gamma_{x\bar{f}^{-1}})$$

$$= \frac{4\pi^3 \hat{\chi}(\bar{g})g}{-y^2 N^5} \sum_{x \in (\mathfrak{o}/N\mathfrak{o})^*} \bar{\chi}(\bar{x}) Q(\gamma_{\frac{x}{N}}) = \frac{4\pi^3 |\mu_\kappa| \hat{\chi}(\bar{g})g}{-y^2 N^5} \sum_{x \in (\mathfrak{o}/N\mathfrak{o})^*/\mu_\kappa} \bar{\chi}(\bar{x}) Q(\gamma_{\frac{x}{N}}).$$

The j-invariant $j(E)$ is known to be real and to generate the Hilbert class field of κ. Since κ has class number 1, $j(E) \in \mathbf{Q}$ so we can choose a model $E_{\mathbf{Q}}$ defined over $\mathbf{Q}$. Deuring's theory associates to $E_{\mathbf{Q}}$ a Grössencharakter $\chi^{\text{Gröss}}$ of κ with values in κ^*,

$$\chi^{\text{Gröss}}(\mathfrak{p}) = \bar{x}\chi(x), \qquad \text{where} \quad (x) = \mathfrak{p} \nmid N.$$

The character χ takes values in μ_κ and $\chi(\bar{x}) = \bar{\chi}(x)$ (this follows from theorem 10, p. 140 of Lang [4], which implies $\chi^{\text{Gröss}}(\bar{\mathfrak{p}}) = \overline{\chi^{\text{Gröss}}(\bar{\mathfrak{p}})}$), so we can rewrite (9.8)

$$(9.9) \qquad L(2, \chi^{\text{Gröss}}) = \frac{4\pi^3 |\mu_\kappa| \hat{\chi}(\bar{g})g}{-y^2 N^5} Q\left(\sum_{(\mathfrak{o}/N\mathfrak{o})^*/\mu_\kappa} S_{x\bar{\chi}(x)/N} \right).$$

The ray class group $(\mathfrak{o}/N\mathfrak{o})^*/\mu_\kappa$ acts on points of order N by

$$x \cdot \left(\frac{y}{N} \right) = \frac{x^{-1}\chi(x)y}{N}.$$

and conjugation acts on these points in the natural way. The cycle

(9.10)
$$U := \sum_{(\mathfrak{o}/N\mathfrak{o})^*/\mu_\kappa} \gamma_{x\bar{\chi}(x)/N}$$

is invariant under both these actions, and hence is defined over **Q**. We get

Theorem (9.11)   *With notations as above, let $\chi^{\text{Gröss}}$ be the Grössencharakter associated to $E_\mathbf{Q}$. Let f be a generator for the conductor ideal and let $fg = N \in \mathbf{Z}$, $g \in \mathfrak{o}$. Then there exists a cycle U on $E \times \mathbf{P}^1 \times \mathbf{P}^1$ relative to $E \times \#$ and defined over $\mathbf{Q}$ such that*

$$L(2,\chi^{\text{Gröss}}) = \frac{4\pi^3|\mu_\kappa|\hat{\chi}(\bar{g})g}{-y^2N^5}Q(U).$$

Remarks (9.12)   (i) $L(s,\chi^{\text{Gröss}})$ is related to the zeta function of $E_\mathbf{Q}$, $\zeta_{E_\mathbf{Q}}$ (defined up to a finite number of factors by the product over all $p \in \mathbf{Z}$ such that E has non-degenerate reduction mod p of the zeta function of the corresponding curve over $\mathbf{F}_p$), by

$$\zeta_{E_\mathbf{Q}}(s) = \zeta_\mathbf{Q}(s)\zeta_\mathbf{Q}(s-1)L(s,\chi^{\text{Gröss}})^{-1}.$$

(ii) Various renormalizations are possible and perhaps desirable to simplify the right-hand constant in (9.11). Dick Gross suggests that the "corrected" formula would give $L'(0,\chi^{\text{Gröss}}) \in Q(U) \cdot \mathbf{Q}$. Since $L'(0,\chi^{\text{Gröss}}) \in \pi^{-2}L(2,\chi^{\text{Gröss}}) \cdot \mathbf{Q}$, one might take $\bar{Q} = 2\pi Q$ (this amounts to modifying (8.8.1) by 2π). A standard Gauss sum calculation (cf. for example Lang [4], p. 289) yields

$$|\hat{\chi}(\bar{g})| = \frac{\sqrt{f\bar{f}}}{N}, \quad \text{whence } |\hat{\chi}(\bar{g}) \cdot \bar{g}|\frac{\sqrt{f\bar{f}g\bar{g}}}{N} = 1.$$

Also for $\tau = x + iy \in k$, we have $y^2 \in \mathbf{Q}$, so

$$L'(0,\chi^{\text{Gröss}}) \in \zeta \cdot \bar{Q}(U) \cdot \mathbf{Q},$$

where ζ is a root of 1.

References for Lecture 9

[1] S. Bloch, Applications of the dilogarithm function in algebraic K-theory and algebraic geometry, pp. 103–114 in *Proceedings of the International Symposium on Algebraic Geometry (Kyoto Univ., Kyoto, 1977)*, Kinokuniyo Book Store, Tokyo (1978).

[2] S. Bloch, *Higher Regulators, Algebraic K-Theory, and Zeta Functions of Elliptic Curves*, Lectures given at the University of California, Irvine (1978). [CRM Monograph Series, no. 11, American Mathematical Society, Providence, R.I., 2000.]

[3] A. Borel, Cohomologie de SL$_n$ et valeurs de fonctions zeta aux points entiers, *Ann. Scuola Norm. Sup. Pisa Cl. Sci. (4)*, **4** (1977), 613–636; errata, **7** (1980), 373.

[4] S. Lang, *Elliptic Functions*, Addison-Wesley, Reading, Mass. (1973). [Second edition: Springer, New York, 1987.]

Bibliography

Artin, M. and B. Mazur, Formal groups arising from algebraic varieties, *Ann. Sci. École Norm. Sup. (4)*, **10** (1977), 87–132.

Atiyah, M. and F. Hirzebruch, Analytic cycles on complex manifolds, *Topology*, **1** (1962) 25–45.

Bass, H. *Algebraic K-Theory*, Benjamin, New York (1968).

Bass, H. and J. Tate, The Milnor ring of a global field, in *Algebraic K-Theory II*, Lecture Notes in Math., no. 342, Springer, Berlin (1973).

Beauville, A. Variétés de Prym et jacobiennes intermédiaires, *Ann. Sci. Ecole Norm. Sup. (4)*, **10** (1977), 304–391.

———. Surfaces algébriques complexes, *Astérisque*, **59** (1978),

Bloch, S. K_2 and algebraic cycles, *Ann. of Math. (2)*, **99** (1974), 349–379.

———. Torsion algebraic cycles, K_2, and Brauer groups of function fields, *Bull. Amer. Math. Soc.*, **80** (1974), 941–945.

———. K_2 of Artinian **Q**-algebras with application to algebraic cycles, *Comm. Algebra*, **3** (1975), 405–428.

———. An example in the theory of algebraic cycles, pp. 1–29 in *Algebraic K-Theory*, Lecture Notes in Math., no. 551, Springer, Berlin (1976).

———. Some elementary theorems about algebraic cycles on abelian varieties, *Invent. Math.*, **37** (1976), 215–228.

———. Algebraic K-theory and crystalline cohomology, *Inst. Hautes Études Sci. Publ. Math.*, no. 47 (1977), 187–268 (1978).

———. Applications of the dilogarithm function in algebraic K-theory and algebraic geometry, pp. 103–114 in *Proceedings of the International Symposium on Algebraic Geometry (Kyoto Univ., Kyoto, 1977)*, Kinokuniya Book Store, Tokyo (1978).

———. *Higher Regulators, Algebraic K-Theory, and Zeta Functions of Elliptic Curves*, Lectures given at the University of California, Irvine (1978).

[CRM Monograph Series, no. 11, American Mathematical Society, Providence, R.I., 2000.]

———. Some formulas pertaining to the K-theory of commutative group schemes, *J. Algebra*, **53** (1978), 304–326.

———. Torsion algebraic cycles and a theorem of Roitman, *Compositio Math.*, **39** (1979), 107–127.

Bloch, S., A. Kas, and D. Lieberman, Zero cycles on surfaces with $P_g = 0$, *Compositio Math.*, **33** (1976), 135–145,

Bloch, S. and J. P. Murre, On the Chow groups of certain types of Fano threefolds, *Compositio Math.*, **39** (1979), 47–105.

Bloch, S. and A. Ogus, Gersten's conjecture and the homology of schemes, *Ann. Sci. École Norm. Sup. (4)*, **7** (1974), 181–201 (1975).

Borel, A. Cohomologie de SL_n et valeurs de fonctions zeta aux points entiers, *Ann. Scuola Norm. Sup. Pisa Cl. Sci. (4)*, **4** (1977), 613–636; errata, **7** (1980), 373.

Châtelet, F. Points rationnels sur certaines courbes et surfaces cubiques, *Enseignement Math. (2)*, **5** (1959), 153–170 (1960).

Chevalley, C. et al., Anneaux de Chow et applications, *Séminaire C. Chevalley*, 2e année, Sécr. Math. Paris (1958).

Chow, W. L. On equivalence classes of cycles in an algebraic variety, *Ann. of Math. (2)*, **64**, 450–479 (1956).

Clemens, C. H. and P. A. Griffiths, The intermediate Jacobian of the cubic threefold, *Ann. of Math. (2)*, **95** (1972), 281–356.

Colliot-Thélène, J.-L. and J.-J. Sansuc, Series of notes on rational varieties and groups of multiplicative type, *C. R. Acad. Sci. Paris Ser. A-B*, **282** (1976), A1113–A1116; **284** (1977), A967–A970; **284** (1977), A1215–A1218; **287** (1978), A449–A452.

———, La R-équivalence sur les tores, *Ann. Sci. École Norm. Sup. (4)*, **10** (1977), 175–229.

Colliot-Thélène, J.-L. and D. Coray, L'équivalence rationnelle sur les points fermés des surfaces rationnelles fibrées en coniques, *Compositio Math.*, bf 39 (1979), 301–332.

Deligne, P. Théorie de Hodge. I, pp. 425–430 in *Actes du Congrès International des Mathématiciens (Nice, 1970)*, vol. 1, Gauthier-Villars, Paris (1971).

———. Théorie de Hodge III, *Inst. Hautes Etudes Sci. Publ. Math.*, no. 44 (1974), 5–77.

———. Poids dans la cohomologie des variétés algébriques, pp. 79–85 in *Proceedings of the International Congress of Mathematicians (Vancouver, B.C., 1974)*, vol. 1 (1975).

————. *Cohomologie étale* (SGA $4\frac{1}{2}$), Lecture Notes in Math., no. 569, Springer, Berlin (1977).

Elman, R. and T. Y. Lam, On the quaternion symbol homomorphism $g_F\colon k_2(F) \to B(F)$, in *Algebraic K-Theory II*, Lecture Notes in Math., no. 342, Springer, Berlin (1973).

Fatemi, T. L'equivalence rationelle des zéro cycles sur les surfaces algebriques complexes a cup product surjectif, These du 3e cycle, Université de Paris VII (1979).

Fulton, W. Rational equivalence on singular varieties, *Inst. Hautes Études Sci. Publ. Math.*, no. 45 (1975), 147–167.

Fulton, W. and R. MacPherson, Intersecting cycles on an algebraic variety, Aarhus Universitet Preprint Series, no. 14 (1976). [Pp. 179–197 in *Real and complex singularities (Proc. Ninth Nordic Summer School/NAVF Sympos. Math., Oslo, 1976)*, Sijthoff and Noordhoff, Alphen aan den Rijn (1977).]

Gersten, S. M. Some exact sequences in the higher K-theory of rings, pp. 211–243 in *Algebraic K-Theory I*, Lecture Notes in Math., no. 341, Springer, Berlin (1973).

Gillet, H. Applications of algebraic K-theory to intersection theory, Thesis, Harvard (1978).

Griffiths, P. On the periods of certain rational integrals. I, II, *Ann. of Math. (2)*, **90** (1969), 460–495; **90** (1969), 496–541.

Griffiths, P. and J. Harris, *Principles of Algebraic Geometry*, Wiley, New York (1978). [Reprinted 1994.]

Grothendieck, A. Le groupe de Brauer I, II, III, pp. 46–188 in *Dix exposés sur la cohomologie des schémas*, North Holland, Amsterdam (1968).

Grothendieck, A. et al., *Théorie des intersections et théorème de Riemann–Roch* (SGA 6), Lecture Notes in Math., no. 225, Springer, Berlin (1971).

Hartshorne, R. On the de Rham cohomology of algebraic varieties, *Inst. Hautes Études Sci. Publ. Math.*, no. 45 (1975), 5–99.

Inose, H. and M. Mizukami, Rational equivalence of 0-cycles on some surfaces of general type with $p_g = 0$, *Math. Ann.*, **244** (1979), no. 3, 205–217.

Kleiman, S. Algebraic cycles and the Weil conjectures, pp. 359–386 in *Dix exposés sur la cohomologie des schémas*, North Holland, Amsterdam (1968).

Lam, T. Y. *The Algebraic Theory of Quadratic Forms*, W. A. Benjamin, Reading, Mass. (1973). [Revised second printing, 1980. See also *Introduction to Quadratic Forms over Fields*, American Mathematical Society, Providence, R.I., 2005.]

Lang, S. *Abelian Varieties*, Interscience Publishers (1959). [Reprinted Springer, 1983.]

————. *Elliptic Functions*, Addison-Wesley, Reading, Mass. (1973). [Second edition: Springer, New York, 1987.]

Lieberman, D. I. Higher Picard varieties, *Amer. J. Math.*, **90** (1968), 1165–1199.

Manin, Yu. Le groupe de Brauer–Grothendieck en géométrie diophantienne, pp. 401–411 in *Actes du Congrès International Mathématiciens (Nice, 1970)*, vol. 1, Gauthier-Villars, Paris (1971).

————. *Cubic Forms*, North Holland, Amsterdam (1974). [Second edition, 1986.]

Mattuck, A. Ruled surfaces and the Albanese mapping, *Bull. Amer. Math. Soc.*, **75** (1969), 776–779.

————. On the symmetric product of a rational surface, *Proc. Amer. Math. Soc.*, **21** (1969), 683–688.

Milnor, J. Algebraic K-theory and quadratic forms, *Invent. Math.*, **9** (1970), 318–344.

————. *Introduction to Algebraic K-Theory*, Annals of Mathematics Studies, vol. 72, Princeton University Press, Princeton, N.J. (1971).

Mumford, D. Rational equivalence of zero-cycles on surfaces, *J. Math. Kyoto Univ.*, **9** (1968), 195–204,

Murre, J. P. Algebraic equivalence modulo rational equivalence on a cubic threefold, *Compositio Math.*, **25** (1972), 161–206.

Nakayama, T. Cohomology of class field theory and tensor product modules I, *Ann. of Math. (2)*, **65** (1957), 255–267.

Quillen, D. Higher algebraic K-theory. I, in *Algebraic K-Theory I*, Lecture Notes in Math., no. 341, Springer, Berlin (1973).

Quillen, D. and D. Grayson, Higher algebraic K-theory. II, pp. 217–240 in *Algebraic K-Theory*, Lecture Notes in Math., no. 551, Springer, Berlin (1976).

Roitman, A. A. Γ-equivalence of zero-dimensional cycles (in Russian), *Mat. Sb. (N.S.)*, **86 (128)** (1971), 557–570. [Translation: Math USSR-Sb., **15** (1971), 555–567.]

————. Rational equivalence of zero-dimensional cycles (in Russian), *Mat. Sb. (N.S.)*, **89 (131)** (1972), 569–585, 671. [Translation: Math. USSR-Sb., **18** (1974), 571–588.]

Rosenlicht, M. Generalized Jacobian varieties, *Ann. of Math.*, **59** (1954), 505–530.

Samuel, P. Rational equivalence of arbitrary cycles, *Amer. J. Math.*, **78** (1956), 383–400.

Serre, J.-P. Sur la topologie des variétés algebriques en caractéristique p, pp. 24–53 in *Symposium Internacional de Topologia Algebraica*, Universidad Nacional Autónoma de Mexico and UNESCO, Mexico City (1958).

———. *Groupes algebriques et corps de classes*, Hermann, Paris (1959). [Second edition, 1975; reprinted, 1984.]

———. *Algèbre locale. Multiplicités*, Lecture Notes in Math., no. 11, Springer, Berlin (1965).

———. *Corps Locaux*, second edition, Hermann, Paris (1968). [Translation: *Local Fields*, Springer, New York, 1979.]

Sherman, C. K-cohomology of regular schemes, *Comm. Algebra*, **7** (1979), 999–1027.

Stienstra, J. Deformations of the second Chow group, Thesis, Utrecht (1978).

———. The formal completion of the second Chow group: a K-theoretic approach, pp. 149–168 in *Journée de Géométrie Algebrique de Rennes (Rennes, 1978), Vol. II, Asterisque*, **64** (1979).

Swan, R. *Algebraic K-Theory*, Lecture Notes in Math., no. 76, Springer, Berlin (1968).

Tate, J. Relations between K_2 and Galois cohomology, *Invent. Math.*, **36** (1976), 257–274.

Tennison, B. R. On the quartic threefold, *Proc. London Math. Soc. (3)*, **29** (1974), 714–734.

Tyurin, A. N. Five lectures on three-dimensional varieties (in Russian), *Uspehi Mat. Nauk*, **27** (1972), no. 5, (167) 3–50. [Translation: Russian Math. Surveys, **27** (1972), no. 5, 1–53.]

Verdier, J.-L. Dualité dans la cohomologie des espaces localement compacts, *Séminaire Bourbaki, Vol. 9*, exposé no. 300 (1965), 337–349. [Reprinted Société Mathématique de France, Paris, 1995.]

———. A duality theorem in the etale cohomology of schemes, pp. 184–198 in *Proceedings of a conference on local fields (Driebergen, 1966)*, Springer, Berlin (1967).

Wallace, A. *Homology Theory on Algebraic Varieties*, Pergamon Press, New York (1958).

Weil, A. *Foundations of Algebraic Geometry*, A.M.S. Colloquium Publications, vol. 29, American Mathematical Society, Providence, R.I. (1946).

———. *Courbes Algébriques et Variétés Abéliennes*, Hermann, Paris (1971).

Index

Printed in the United States
By Bookmasters